Praise for

The Tao of Alibaba

"The explosion of the Chinese economy over the past forty years has been an earth-shaking event. Wise government policies triggered it, and a dynamic private sector drove it. Brian Wong has done an absolutely brilliant job of drawing out the special genius of Jack Ma and the *Tao of Alibaba*. This captivating tale will fundamentally change American perceptions of the China story. An absolute must-read!"

—Kishore Mahbubani, founding dean of the
Lee Kuan Yew School of Public Policy,
NUS, and author of *Has China Won?*

"This should be required reading for business school! *The Tao of Alibaba* bridges East and West and provides a fresh alternative to Western frameworks for leadership and company building. Wong boils down Alibaba's culture to a systematic, clear, and repeatable process. For the first time, founders around the world have the complete recipe to Alibaba's secret sauce."

—Connie Chan, General Partner, Andreessen Horowitz

"People in the West should resist the temptation to discount tech firms like Alibaba just because of how different the operating environment is in China. But being Chinese is core to their corporate DNA. No one is better placed to unpack Alibaba's secret sauce than American-bred and educated Wong, who was at founder Jack Ma's side for years. His insights into what makes Jack tick and what makes Alibaba great are invaluable to anyone interested in starting and scaling a tech business."

—Geoffrey Garrett, dean, Robert R. Dockson Dean's Chair
in Business Administration and professor of management
and organization, USC Marshall School of Business

"An insider's account of one of the world's most important companies and a rare opportunity to see not only what makes Alibaba tick, but also how it helped change China."

—Peter Cappelli, George W. Taylor Professor of Management, the Wharton School, and coauthor of *Fortune Makers*

the
Tao *of*
Alibaba

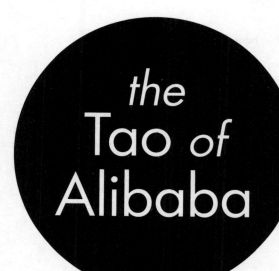

the
Tao of
Alibaba

*Inside
the Chinese
Digital Giant that Is
Changing the World*

Brian A. Wong

PUBLICAFFAIRS

NEW YORK

PublicAffairs
Hachette Book Group
1290 Avenue of the Americas, New York, NY 10104
www.publicaffairsbooks.com
@Public_Affairs

Printed in the United States of America

First Edition: November 2022

Published by PublicAffairs, an imprint of Perseus Books, LLC, a subsidiary of Hachette Book Group, Inc. The PublicAffairs name and logo is a trademark of the Hachette Book Group.

The Hachette Speakers Bureau provides a wide range of authors for speaking events. To find out more, go to www.hachettespeakersbureau.com or call (866) 376-6591.

The publisher is not responsible for websites (or their content) that are not owned by the publisher.

Library of Congress Cataloging-in-Publication Data
Names: Wong, Brian A., author.
Title: The tao of Alibaba : inside the Chinese digital giant that is changing the world / Brian A. Wong.
Description: First edition. | New York : PublicAffairs, [2022] | Includes bibliographical references and index.
Identifiers: LCCN 2022019131 | ISBN 9781541701656 (hardcover) | ISBN 9781541701663 (ebook)
Subjects: LCSH: Alibaba (Firm) | Electronic commerce—China. | Internet marketing—China.
Classification: LCC HF5548.325.C6 W65 2022 | DDC 381/.1420951—dc23/eng/ 20220428
LC record available at https://lccn.loc.gov/2022019131

ISBNs: 9781541701656 (hardcover), 9781541701663 (ebook)

LSC-C

Printing 1, 2022

To my mother, Barbara Jean Wong,
who taught me the importance of a generous heart.

Contents

Introduction

From the time that I began writing this book two years ago up till its completion, the world has changed in ways we never imagined. The global pandemic has incurred psychological and physiological trauma upon the world that will last a lifetime. Climate change has become undeniable as breaking news stories on unprecedented cyclone bombs and shocking melting polar ice caps have become regular news stories. The war in Ukraine and US-China relations have exposed geopolitical fault lines reigniting concerns of a new cold war. In short, a myriad of existential crises are now facing not just one or a few countries but all of humanity.

Jack Ma, Alibaba's founder, constantly reminded us "where most people complain is where the opportunity is...and where there is trouble...there is opportunity." If that's the case then one could say that today there is a massive opportunity before all of us. And what it requires is a major mind-set shift in thinking about how we can solve these existential crises using the tools at our disposal.

In writing this book, I set out to make the case that today's unique technological era offers us the potential to become part of the *solution* to many of today's global problems and that the ethos Alibaba has developed and shared is the road map. Seeing is believing, and Alibaba has shown time after time how it can be done.

This is a book about a company, Alibaba, that not just shaped a huge industry, e-commerce, but also impacted an entire country, China, on its digital journey that began with few if any of the attributes most say are necessary for a typical successful start-up. What it did have was Jack Ma, an unconventional, visionary entrepreneur who has described his unlikely path as being like a "blind man riding on the back of a blind tiger." Asked once during a conference at Harvard about the secret to Alibaba's spectacular rise to dominance, Jack, who loves to sound outrageous, said: "Alibaba succeeded because we (1) had no plan, (2) no technology, and (3) no money."

But Alibaba had what turned out to be more powerful guiding attributes—a truly deep sense of purpose, a social equity mission, and a focus on spreading a new type of inclusive, shared wealth, all built on a management foundation that placed the importance of values above key performance indicators (KPIs). Its founder had an intuitive understanding of the company's first base of customers, the underserved small and medium-sized businesses and entrepreneurs in need of help. Given the right encouragement and education from Alibaba, those ambitious entrepreneurs, Jack knew, would seize the opportunities in the vast new frontier of 1s and 0s.

Alibaba's digital model both benefited from the remarkable dynamism of China's economy—one of the most far-reaching explosions of prosperity in history following the launch of the government's economic reform programs in the 1980s—and energized it. At the same time, Alibaba's rapidly expanding platform helped to spread prosperity to overlooked regions and businesses in new and transformational ways.

I was the first American and the fifty-second employee of this unusual company, spending countless hours toiling, initially beside my colleagues on unmatched sets of chairs and couches in Jack's apartment in the provincial city of Hangzhou, and saw it struggle through its early years, even a near bankruptcy, and I can assure you that Jack, though delighting in sounding irreverent, was using only a little hyperbole in his descriptions of our early days of no plan, no technology, and no money.

But I grew to admire and embrace his mission and, especially, the power of the distinctive model that he articulated. It's a model that is replicable and can be employed by leaders or organizations to make them better, more efficient, and more socially impactful.

Those tenets, more of the heart than the head, informed the agile management systems we developed to constantly reinforce the Alibaba culture: it was our "secret sauce." Alibaba's leadership team was extremely inventive and flexible in implementing those values and using them to develop the strategies that made our highly motivated workforce so productive and resilient.

That mix of qualities made up what we called "the Tao of Alibaba," which involved finding balance and harmony even in circumstances filled with contradictions. In this book I describe how they worked in practice and how they can be used to drive an entrepreneurial enterprise.

In the following chapters, I explain how the organization that Jack and the seventeen other founders built launched a new development paradigm not just for creating wealth but also for addressing poverty and job creation, in addition to inspiring the growth of a vast new marketplace. These lofty results can be traced back to the guiding light of Alibaba's mission, vision, and values, which gave direction and clarity to the hundreds of thousands of staff who made our digital ecosystem flourish.

It is a model my team and I helped codify in the programs used, first, to teach our systems to the incoming Alibaba managers leading our global expansion and, then, as part of Jack Ma's passionate social mission to encourage and guide young entrepreneurs in emerging markets as part of a development paradigm that is changing economies in Africa, Southeast Asia, and other emerging regions. These are the sorts of lessons you won't find in standard MBA courses. They are observations and conclusions that come from years of trial and error and reflection about the principles that guided the decision-making of Alibaba's team amid a market that had no precedent or road map for success. In short, the Tao of Alibaba approach enables a mind-set and way of thinking that liberates a group of individuals to rigorously and effectively pursue a common cause they all believe in.

Our company mission was "to make it easy to do business any-where." But Jack's vision was also unique because the aim was to apply this formula largely to outsiders—small businesses, entrepreneurs, and long neglected regions of China—and provide them unimagined op-portunities. Jack and some of his lieutenants loved Chinese martial arts novels, so early on we used some of the legendary characters from those *wuxia* stories to emphasize our self-image as scrappy underdogs who placed ethical values above the lust for power as we battled on behalf of those left behind or oppressed. The message was that we would succeed only by staying true to our principles. My team members and I were part of a cause, not just holding jobs.

These values, which evolved over the years, were articulated in 2019 as the following:

1. **Customers first, employees second, shareholders third.**
2. **Trust makes everything simple.**
3. **Change is the only constant.**
4. **Today's best performance is tomorrow's baseline.**
5. **If not now, when? If not me, who?**
6. **Live seriously, work happily.**

In Part I of the book, "The Alibaba Journey," I explain Alibaba's origins and how I stumbled into this remarkable adventure and how we steadily transformed a relatively simple business-to-business e-commerce platform, piece by piece, into a highly competitive global ecosystem. That structure assisted our small and medium-sized business clients and consumers by providing access to buyers and sellers, finance and pay-ments systems, cloud computing, and logistical support.

Part II, "The Secret Sauce," unpacks the philosophy, management, and leadership practices that form the foundation of Alibaba's success, distilled into lessons and usable frameworks for entrepreneurs and busi-ness leaders. This section demystifies this highly successful, replicable set of management strategies and practices. In Chinese, the word *tao* (道) translates to "way," as in the nature or essence of something. The Tao of Alibaba is exactly that—a consciously cultivated ethos and spirit that infuses all of the strategies and management systems. I end this part with an exploration of Alibaba's leadership practices.

In Part III, "The New Digital Frontier," I examine Alibaba's impact—within China and across the developing world—and how the company has helped guide the transition to a more inclusive digital-first economy in emerging markets. Through the curriculum and fellowship programs we created, we taught and helped apply the techniques from this paradigm so that entrepreneurs could leap over traditional barriers to growth and fulfill the development potential of countries and regions.

In the past few years, the environment has shifted. Alibaba has been hit hard by changing government policies. China has benefited greatly from the success of the digital economy, but regulators have sought to balance a variety of policy concerns that have arisen with the emergence of these giant companies and their founders. A broad-based tightening up of measures has affected Alibaba and many other digital giants. In particular, the initial public offering (IPO) of Ant Group, a financial technology spin-off from Alibaba—which, at $37 billion, was slated to be the world's largest IPO—was abruptly halted due to regulatory concerns. This event captured headlines throughout the world and marked a turning point in how big tech was to be governed in China.

China is not alone in this reassessment of the power of the major technology companies. The problems include the potential for anticompetitive behavior, insufficient consumer protection, data privacy concerns, and the impact of so-called surveillance capitalism. Governments in the United States and the EU, as well as China, are investigating suspected abuses and reportedly exploring strategies for reducing the dominance of these companies, from imposing new regulations to possibly breaking them up. These concerns have been focused on the dominant social media, search, and e-commerce giants, such as Meta Platforms (formerly Facebook), Amazon, and Alphabet (parent company of Google), as well as Chinese giants such as Tencent, Kuaishou, Bytedance, and also Alibaba. In addition, rising tensions between the United States and China over competition in areas ranging from regional power, cyberattacks, and corporate espionage to trade in goods and services, as well as human rights issues, have created even greater uncertainty about the environment in which the big technology companies operate.

While these policy issues have clouded the prospects of many large internet companies, the situation continues to evolve. And no matter how these challenges are resolved, and whatever kind of Alibaba will emerge from the crackdown, the distinctive digital model for growth that the company designed remains a powerful tool for achieving a new type of inclusive success.

My hope in writing this book is that entrepreneurs, academics, and policy makers with lofty aspirations to solve society's problems will gain inspiration and guidance from the lessons distilled from the Alibaba experience. I hope that instead of listening to the voices that give them one thousand and one reasons why something is not possible or why it will fail, they instead look to the fascinating story of an individual and a company that started with nothing from the most unlikely of places to become a powerhouse that has shaped the internet era. What follows is a road map to replicating and delivering that success.

Part I

The Alibaba Journey

1

Early Days

Open Sesame

All during my childhood in Northern California, China hung like a lantern in the distance. I knew it was the place where my ancestors came from, but it had little to do with my daily life. Yet at the same time, much of what I knew was cast in its glow.

I grew up in a thoroughly American immigrant family of strivers. My dad was the fourth of nine children and the first in his family to be born on American soil after his parents arrived from southern China. They placed enormous value on educational attainment and community service, which led my father and several uncles to become physicians. My mom's family came to the United States even earlier, and her parents were farmers, settling near Sacramento, where row upon row of fruit trees—apples, apricots, cherries, and peaches—sat amid beautiful rolling hills, which I loved to visit as a child.

Growing up in Palo Alto was a pleasant typical suburban childhood for a kid: bicycling every day to school, participating in the local soccer league, attending church on the weekends, and never doubting that I, too, would become a doctor. But China, in the form of Chinese values and culture, was also as much a part of my identity as my favorite San Francisco Giants cap. Every year we gathered with my dad's side of the family for a big reunion, usually hosted at my aunt Maegan's house in Seattle, and it became a tradition for my dad to use the occasion to share with us his thoughts on Chinese values, such as filial piety and the importance of education. It was meaningful but in truth, as the years passed, a relatively slender thread connecting me to my heritage. During middle school that would change.

Familiar Photos in a Family Temple

In 1985, my dad was invited to give a talk at China's first international ophthalmology conference in the southern city of Guangzhou. I was in sixth grade, and the conference happened to fall during my winter break, so he invited me to join him. I leaped at the opportunity, especially since we planned to visit his ancestral village after the conference.

Once the conference had ended, my excitement built as we climbed into a rented car and headed out from the gleaming modern city of Guangzhou into the countryside to a village in Taishan, where my paternal grandfather was born. It was a journey back in time to a far different China than the one experienced in the hotel conference spaces. The highway eventually gave way to a bumpy dirt track filled with a neverending stream of battered cars, overburdened motorcycles, and bicycles. It was the first time I saw three-wheeled rickshaws, most powered by sputtering engines, lurching along under huge mounds of tightly packed bags and even cages with clucking chickens.

As we finally bounced our way onto the streets of Taishan, we were greeted by family members. We followed a family elder from one small house to another while he introduced us to smiling relatives and neighbors. My dad, who had learned the family's local Chinese dialect as a child, chatted amiably, but I understood little and was quiet.

Later, as we walked through the village, I had my first encounter with rural China. It was an entirely new world for me, the first time I had strayed so far from my middle-class American comfort zone. The dirt roads were dusty and uneven, stray dogs roamed about, and all around us were rice paddies on which the villagers toiled to feed their families. Young boys and girls played under the afternoon sun, bundled in tattered hand-me-down sweaters and trousers. It was jarring to my young mind, and I wondered what it might take to improve what to me felt like extremely modest circumstances.

But, despite the striking contrast with the living conditions I was used to, in their faces I saw the faces of my cousins in California. I imagined *myself* playing among the village children and imagined what my childhood might have been like had my grandfather never summoned the courage to make his way to America decades before.

As I marveled at the vast breadth of the world for the first time, the riddles of opportunity and fate, I struggled with conflicting emotions. I felt both a sense of belonging and distance. In that moment, I realized there was little about their lives or experience that I could understand or relate to, but I resolved that I would change that.

I decided then and there that one day I would return to China.

The Road to Nanjing

When it came time for college, I chose to attend Swarthmore College, outside Philadelphia. I was still certain that I was going to become a doctor, like my father, and I combined premedical and literature studies with courses in Chinese history and language. Following my first visit to my family village, I started attending Chinese language classes back home in Palo Alto, beginning my progress toward fluency, and made it a point to visit China during my college summers as a language student in Peking University and as a volunteer English teacher in Yunnan province.

As I neared graduation from Swarthmore, I felt restless to see something of the world and made a decision that I was not ready yet to begin my medical studies. Thinking about what else I might do, one of my language professors suggested that I might consider applying to a one-year program at the Hopkins-Nanjing Center, an institute jointly run by Johns Hopkins University's School of Advanced International Studies and Nanjing University. I did not hesitate. That became my priority.

I moved to Nanjing in the fall of 1996 guided by a plan. I had decided to research Chinese health-care access, which aligned with my ambition to work as a doctor in the developing world. This path, I felt, would provide a solid career trajectory and also allow me to satisfy my strong personal desire to contribute to the common good, lifting up poor individuals and families through improved health. China, it seemed, offered an ideal opportunity to honor and fulfill my commitment to supporting individuals and communities and to learn how better health care could raise living standards. I would also observe how this development model was being practiced in an emerging market like China.

The China of the 1990s, while still relatively poor, was a world undergoing extraordinary change. The government's economic reforms, launched in the late 1970s, had started to transform urban centers with a proliferation of private as well as state enterprises focused on developing manufacturing and export markets and improving the infrastructure. Workers' wages were a fraction of those in the West, but they were rising and generous by Chinese standards of the time. They were providing a new level of prosperity to millions who had enjoyed little before.

The longer I lived in Nanjing and traveled around the country the clearer it became that, for a developing country such as China, the business boom seemed to be doing more to rapidly lift living standards and the quality of life for average Chinese than the health-care system. Rising wages were transforming workers into consumers, suddenly eager to snap up popular brands filling store shelves. Education was improving. Modern housing was springing up. Simply eating at a fast-food chain like KFC or McDonald's exposed customers to service and hygiene standards that were significantly higher than the traditional Chinese restaurant. These Western restaurants provided stations for handwashing before dining and standardized systems for food safety. At the same time, Chinese household habits and everyday hygiene practices were improving.

These observations hit me like revelations. I had always regarded commerce with a skeptical eye. With my limited experience, I saw the business world as one dimensional: it meant little more than money-obsessed corporations doing anything for profits. But in China, I was starting to realize, competitive businesses had the capacity to also effect meaningful change in the lives of their customers as well as their workers. A business could be an instrument of social change.

My expanding perspective on economic growth and business was supported by some of the friendships I made during my year in Nanjing. On many of my weekends at Hopkins-Nanjing, I would take the three-hour train ride to visit a new circle of friends in Shanghai who were part of a fascinating world of ambitious entrepreneurs. They worked in fields like technology, media, and finance, often standing at the frontiers of new directions and breakthroughs. I was learning from them how businesses are designed, how they compete and grow, their cultures and what kind of impact they can have.

In the spring, recruiters from various industries came to the Nanjing University campus for job interviews, and after a string of information sessions and coffee chats, I again chose to put off the pursuit of medicine and decided consulting would be an exciting, well-paid way to begin my education in how the business world operated. Soon I received an offer that promised the sort of tutorial I was hungry for—working as a market-entry specialist on the Asia team at a boutique American consultancy. I spoke Chinese, was getting familiar with Asia, and felt good about the opportunity to develop an entirely new skill set.

I traveled constantly, particularly in southern China. I bounded from one bustling new factory town to the next along China's vast coastline—from Foshan to Fuzhou, Shunde, and Shantou. Some of my assignments seemed mundane, but each one opened a new window into China's economic expansion and the growing opportunities.

On one project, I was tasked with determining the country's demand for industrial water pumps, allowing me to explore the entire market, from real estate developers to individual contractors. Another assignment required that I master the industrial procurement process for nylon filaments. I started in Taiwan, where the raw nylon pellets were produced, and continued to the filament extrusion machines, manufactured in southeast China, and then on to the Shanghai workshops of weavers and knitters, who transformed the filament into industrial fabrics.

Our clients were mostly Western multinational corporations mesmerized by the promise of Asian markets, but China, a once sleeping giant, was the ultimate prize because of the size of its population and its markets.

Indiana Jones in a Business Suit

We were pioneers, and our relative youth did not stand in the way of our tackling—some of us, anyway—some very important projects. Being young and energetic, I happily followed assignments from fancy big city hotels to small industrial towns, where I often found myself being ferried on the back of a whining motorcycle, clinging to a briefcase, my necktie flapping wildly in the breeze.

One of my classmates from the Hopkins-Nanjing Center had gone into the health-care field and joined a start-up team that was establishing a joint venture hospital to bring in state-of-the-art Western medical technologies. I joined him on a tour of some well-established Chinese hospitals, where we were struck by the rudimentary, outdated equipment and rundown facilities. They were, in fact, fairly typical for the country at that time.

He and his company quickly engaged in negotiations to obtain overseas funding to finance construction of a modern hospital in Beijing. They were recruiting internationally recognized physicians from abroad, tapping global pipelines of talent and technology to help serve China's needs.

My friend impressed me because he and his team were not just cutting deals but were also improving the lives of countless Chinese. Their success went well beyond treating an illness or curing a sick patient. They were creating a foundation for uplifting communities. Their achievements swept away my lingering apprehensions about the usefulness of commerce and solidified my conviction that business could be a powerful force for development and fulfill my long-held values and commitment to societal improvement.

It was a heady time on perhaps the most wide-open frontier in modern economic development—but, as I soon found out, it wouldn't last.

Exploring Public Service

As often happens, I would discover, crises can come from unexpected sources. An Asian financial crisis was set off in July 1997 by a sudden collapse in the value of the Thai baht. No one appreciated until it was too late how this could trigger a chain reaction of selling in other markets, involving an array of securities and commodities.

Investors all over the world, anxious about the unforeseen devaluation and market reaction, began withdrawing funds from other Southeast Asian economies, pushing some into recession. Commodity prices tanked. Markets that relied on trade in commodities declined, from Russia to Brazil to the United States. As confidence in Asian economies dropped, consulting contracts dried up. My peripatetic life screeched to a halt, and I found myself languishing with no new projects or assignments.

Near the end of 1998, my firm offered me a new position in the San Francisco office, but with the sudden shift in the business climate, I was too focused on work to enjoy the homecoming. I had transferred into our company's technology division. While the first internet or dot-com bubble was about to take off at that time, our firm's focus was more on legacy telecom or established internet service providers. I found myself flying off each week to far-off locations such as Kansas City to serve Sprint or Boston to serve MCI—not Sand Hill Road, the heart of Silicon Valley venture capital firms, in Palo Alto. There was a buzz in the air, and everyone was talking about this thing called "the internet," but it wasn't playing a prominent role in the work I was doing.

As a consultant in China, I had felt like I was plowing rich soil with real promise. Being a consultant in America meant shuttling from one faceless chain hotel to another, giving drab presentations to bored audiences. I had managed to move back from halfway across the world, close to the center of Silicon Valley and tech innovation, yet I was stuck in a sort of corporate dead end of overhead projectors, generic conference rooms, and early cocktails.

Something had to change. By chance, I learned through a college friend that the San Francisco mayor, Willie Brown, was seeking a new ombudsman for the Richmond, Marina, and Chinatown districts. I thought my background might be appropriate, so I applied and was soon meeting with small business owners and the residents. One area of particular interest for me were the interactions and assistance we provided to those from low-income households or in need of various city services and support. I saw firsthand how much of a difference government assistance can make to those in need in a surprisingly wide number of areas such as health services or providing access to housing subsidies. I felt grateful to be in a position that gave me the ability to help provide sorely needed resources.

Ultimately, however, I encountered both the benefits and limitations of local government. Different groups with different priorities compete for scarce resources and attention, and the ensuing infighting can paralyze even the most well-meaning programs. And during a campaign year, there is also the unceasing need for political fundraising. This outsized reliance on money in politics underscored an important distinction between government and private enterprise. Establishing a sustainable business is never easy, but once there is steady cash flow, it allows the leaders to exercise more control over their agendas and priorities.

I admired government's core function of providing for the common good, but I wondered which sector was best equipped to overcome all the barriers to delivering equitable, sustainable prosperity: government or private enterprise. I was wrestling with these challenges when I received an email from Joseph Tsai in the fall of 1999.

I had been introduced to Joe a couple of years earlier at a holiday party organized by a mentor in Hong Kong, and that chance meeting had left a memorable impression. Joe had attended Yale College and Yale Law School, worked at a large law firm, and then joined the Hong Kong office of Investor AB, a prominent Swedish investment firm. He exuded a quiet confidence.

Joe said he was in San Francisco for a few days on a business trip and wondered if I had time to get together. I was interested to hear what he was doing, and so we arranged to meet up at the St. Francis Hotel for a drink. I had recently learned through a mutual friend that Joe had left the investment firm and had taken a surprising leap to a little-known Chinese internet start-up. Joe mentioned that he would be bringing along the company's founder—a business newcomer from the provincial coastal city of Hangzhou. His name was Jack Ma.

A Cave of Treasures

Before meeting, I decided to check out the start-up's website and was amused by its Disney-esque name, Alibaba. I learned later that Jack had felt the "open sesame" incantation from the Arabian folktale (in the story, a poor woodcutter named Ali Baba learns that those magic words unlock a secret thieves' cave filled with treasure) expressed his aspirations to open the doors to fortune for China's emerging small business owners.

At the time, the website did not strike me as an entryway to great riches. I arrived at an extremely plain webpage, covered in text, that merely presented lists of products for sale. There were a few grainy images of mundane items, ranging from hair dryers to garden hoses. Alibaba did not even have a logo.

Given the unsophisticated design of the website, it was clear this was a far cry from the Silicon Valley dot-com ventures that were grabbing headlines and investments—and talent. I recalled that a few months earlier, a friend had told me about a summer internship he would be doing at a company with an even sillier name, Google. This friend had been training to be a doctor at the University of California, San Francisco, Medical School but left that path to try work at the new search engine firm. That was the first time I had heard of Google, but that a promising physician would jump careers to work there encapsulated for me the sense of possibility in the air.

Internet commerce was far less developed in China, but just as Western films, music, household brands, and fashions were flooding into the country, technological innovations were making their way into the Chinese and Asian markets. The most prominent Chinese tech firms were merely internet portals. Yahoo! was the most inspiring American dot-com success story, so much of China's best talent sought to replicate its services for the Chinese language market.

Three such companies established themselves as the leaders: Sina, Sohu, and NetEase. The founders of those companies all had prestigious educational pedigrees and polished track records in software and technology, so they were magnets for the limited amounts of capital being invested in the Chinese tech market.

The Alibaba founder, Jack Ma, could not have been more different. Jack had attended a third-tier Chinese university in a second-tier city and began his career as an English instructor at a modest teacher's college in his home province of Zhejiang. As I walked into the St. Francis to meet him and Joe, I saw a slight, nondescript man wearing a big, confident smile.

Jack spoke fluent English at a furious pace that hinted at his outsized personality. We quickly developed a good rapport as he sipped tea and talked about his aspirations for Alibaba. Within minutes, I was seized by a sense that I was in the presence of an extraordinary thinker. Jack's charisma flowed from his passion and his supreme self-assurance: he talked fervently about his company, which seemed to have little to do with the uninspiring website I had seen earlier. His vision, he told me, was to make Alibaba a world-class platform for conducting online commerce globally. The immediate goal was to grow the company to a million users within one year—though Alibaba had just thirty thousand users at the time. Jack issued this proclamation with such authority that I didn't even think to question it.

When Alibaba first launched that year, Jack's initial goal was to create a centralized, easily searchable directory that brought Chinese businesses and suppliers, from big city concerns to humble small businesses in the provinces, online. He talked not at all of performance metrics, cash flow, or return on investment, the usual start-up buzzwords, but of his mission to give talented, hardworking Chinese entrepreneurs something few had enjoyed before—opportunity. These were his people. He knew them. Give these small manufacturers and suppliers greater access to buyers, he believed, and everyone would prosper. Innovation would flourish.

I agreed it was exactly the kind of service that the Chinese market needed, especially for foreign companies wanting to find opportunities to do business in China. During my days hopping around Asia and attending trade shows, I often had to spend hours scouring telephone-book-sized printed catalogs to find the right contractors and suppliers for my projects. Occasionally, I would get lucky and find a CD where I could download the data, but even then the information was often outdated.

Unlocking Small Business Potential

China's capacity for manufacturing for export markets and selling at attractive prices was already booming at that time; the missing piece was getting around infrastructure and communications restraints and giving these suppliers an even more effective means of connecting efficiently with buyers around China and in other countries. This way, buyers could easily locate the products they needed from the right manufacturers or intermediaries, without the need to trudge around to regional trade shows and hope for the best. Jack understood the promise of such a platform, and he articulated a remarkably clearsighted, positive strategy for the role he wanted Alibaba to play.

Listening to him lay out the plan, Jack appealed to my head and my heart. He expressed not a conventional strategic plan but a vision of how this digital system could support small and medium-sized Chinese businesses and unlock their exceptional potential.

Jack then put down his teacup and asked if I would consider joining them. Fanciful as it sounded, I knew I had to take the offer seriously. The appeal wasn't just Jack's missionary zeal or the clear prescience of his business plan. I was also impressed by the commitment of his newly hired CFO, Joe Tsai. If a high-flying investor like Joe, with his prestigious credentials and careful approach to business, was willing to abandon his soaring career trajectory and tie his fortunes to Alibaba, it was hard to say no.

I didn't make a firm commitment, but in my heart I knew I wanted in. I said I would think about it. I received my offer letter from Alibaba a few weeks later. I became the company's fifty-second employee.

I did not need any more convincing, but it seemed like fate when I heard an encouraging story about this would-be titan of internet commerce. A childhood friend of mine had been working for Goldman Sachs in Hong Kong, at the bank's private equity unit. He was home in

Palo Alto for the holidays, and he stopped by my parent's home where we met to catch up.

His team, he told me, had started investing in Asian tech firms, and he happened to mention one example of a company they were looking at—a Chinese business-to-business e-commerce platform called Alibaba. I could barely contain my excitement. An investment from a major Wall Street firm like Goldman would provide an enormous boost to the company's credibility. At the very least, I thought, it would ensure that the start-up would be sticking around for a while.

I returned to Asia filled with both excitement and a sense that I was truly fulfilling my intertwined passions. I wanted to be involved in China, to bring prosperity to an emerging economy and get involved with emerging technologies in a constructive and impactful way.

After my two years back in the United States, I missed the adrenaline rush of being on the frontier of that raw and rapidly developing part of the world. In San Francisco, I had gotten a taste of public service while witnessing the internet's blossoming and the rising tide of transformative internet companies. In joining Alibaba, I was catching a wave that would spread the internet magic much further, to huge, largely untapped markets. I wanted to help bring some of Silicon Valley's ethos of innovation and entrepreneurship to China and figure out how to bring my commitment to service to this interesting new venture.

Little did I know, it would be China bringing that spirit to me.

2

Alibaba Meets the World

The Rise of China's Digital Economy

On a breezy evening in September 2009, I stood backstage at Hangzhou's leading sports venue, Dragon Stadium, peeking from behind a curtain at thirty-five thousand excited Alibaba employees and their families. It was our ten-year anniversary, but this was most assuredly not a typical stuffy corporate celebration with droning speeches on quarterly performance or half-hearted fist pumping. What the employees got was Jack Ma, in a flowing white wig, lip-synching an Elton John song blaring over the sound system—while wooing a dramatically bashful Joe Tsai, who was in high heels and a shimmering short dress. The audience whooped at the spectacle. It was 100 percent in the Alibaba spirit, with team flags, whistles, and thunder sticks to boot.

Alibaba had arrived. China's digital economy had arrived, floating on a magical cloud of 1s and 0s. The road to China's modernization had been, in fact, little short of miraculous, and Alibaba's meteoric rise was a reflection of that success.

In just ten years, the scrappy business-to-business (B2B) platform, started by an unlikely motley crew in Jack's apartment, had become arguably China's most powerful—and most unconventional—internet firm. We were brash, by design. We cultivated a persona that was decidedly different from other Chinese companies. We thought in terms of giant leaps, not increments. We weren't a monolithic, faceless corporate entity but a highly visible organization with a large personality cast in the image of our highly creative founder, Jack Ma.

By this time, much had changed since that first day I had to step over a mound of smelly shoes just to knock on that Lakeside Gardens apartment door, the humble abode that served as Alibaba's first office location. Gone were the anxious nights I had spent lying awake in my apartment, wondering if I'd made a terrible mistake by turning away from a once promising career, leaving behind my ambitions to become a doctor, a business consultant, or a public servant and joining this untested start-up. Alibaba was no longer untested: in fact, it had been tested many times, teetering on the edge of bankruptcy, and at each turn had proven itself a capable, even essential Chinese tech company. It had been a long road getting to that point.

The Fourth Industrial Revolution

To put the events of that period into context, it helps to zoom out and identify what was happening on a more macro level both within China but also along the continuum of industrial change. China's digital transformation and the explosive expansion of the innovative digital economy are products of what historians describe as a fourth industrial revolution.

In some ways, Alibaba benefited from China having been passed over for several centuries by the rise of the industrial West. Starting in the 1980s, it was playing a frantic game of catchup, struggling to build the infrastructure and commercial spirit needed to drive the development of a modern economy. There were limited legacy systems or legacy constituencies standing in the way of the adoption of digital innovations.

Historians describe global economic development as having gone through a series of industrial revolutions, each led by technological innovations and new business opportunities. The first shift marked the initial transition from agrarian societies into ones led by industrial production and new infrastructure. Starting in the eighteenth century, railroads and steam and hydro power ushered in mechanized production and factories. Electrification led the second revolution. Electric power enabled more rapid and more reliable mechanization in the form of assembly lines and mass production, creating divisions of labor and greater efficiencies.

The invention of computers was the catalyst for the third step, automating production and enterprise management. At the end of the 1960s, computers were first linked, paving the way for the creation of networks—the internet. As computing power grew rapidly, economics, corporate life, and even social life were transformed. The digital economy was taking over.

A Historic Perspective
Navigating the next industrial revolution

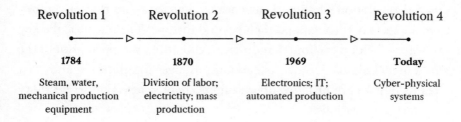

Revolution 1	Revolution 2	Revolution 3	Revolution 4
1784	1870	1969	Today
Steam, water, mechanical production equipment	Division of labor; electrictity; mass production	Electronics; IT; automated production	Cyber-physical systems

Source: World Economic Forum

Characteristics of the four industrial revolutions.

Klaus Schwab, founder of the World Economic Forum, explained the next step in this historic process, the "fourth industrial revolution." He described a new digital revolution built on the pillars of the third, "characterized by a fusion of technologies that is blurring the lines between the physical, digital, and biological spheres."

The core building blocks of this revolution—computer hardware, software, and networks—have been around for decades, but they are now spurring the development of innovative applications that are impacting every aspect of how we conduct business, govern, and engage in all sorts of personal and commercial relationships and even our personal identities. Erik Brynjolfsson and Andrew McAfee, two MIT professors, have christened this "the second machine age," while others have adopted the term *Industry 4.0*.

For years, Jack Ma has described the changes as data technology, or DT, a progression of the information technology (IT) wave that helped launch this new era of the digital economy. The initial IT methods were proprietary: data was generated and analyzed within closed systems. DT supports a different approach to technology in which access to data is open and its use can be applied to help other collaborators and partners. Through these features, the DT revolution is creating entirely new types of businesses, from everyday commerce to biotechnology, renewables to robotics, gene sequencing to quantum computing.

One important difference between the third and fourth industrial revolutions is the vast scale of adoption and the velocity of innovation. As Schwab observed, the cloth-weaving machine spindle—the first industrial revolution's hallmark invention—needed more than a century to make its way through and then beyond Europe. The internet, the key innovation behind the fourth industrial revolution, wrapped around the world in less than a decade. As networks all over the globe expand, vast streams of data are bringing about unprecedented levels of connectivity and fresh applications.

There is no better example of the fourth industrial revolution—both its progress and its promise—than Alibaba. From that small team in Jack's apartment twenty-plus years ago, the company now has more than 200,000 employees and hundreds of billions in market value. Its family of websites and platforms, the Alibaba ecosystem, supports more than ten million merchants, helped create fifty-four million jobs, and continues a climb beyond one billion active users.

The China Jack Ma Found and Changed

Jack Ma was an unusual pioneer in this revolution. He's not a technologist, a coder, an engineer, or even a business school graduate. He runs largely on intuition, and he has an uncanny ability to understand people and their yearnings, a knack for understanding what people need and want. And he has a strong moral compass that points him toward wanting to support societal change and economic inclusiveness. Those may not sound like the best digital credentials, but, as I will explain, his story makes clear why they have made all the difference.

His journey, well documented elsewhere, began with an unexpected adventure during his first visit to the United States in 1994 on business. Because of his facility with English, he had been hired to assist a Chinese state-owned enterprise collect a debt in California from the company's business partner. While in the United States, knowing very few people there, he went to Seattle to visit a friend's family, and by chance the son-in-law of the friend introduced Jack to this new phenomenon, the internet.

It was Jack's very first online experience. He was entranced and was initially afraid to even touch the keyboard, he recalls, fearing he might break it. He was persuaded it was OK and was urged to use the search browser to look something up, so Jack thought for a moment and then typed in "beer" and "China," seeing what Chinese brands would pop up. The search engine responded with nothing, not even the export brand Tsingtao, a telling moment. Rather than being frustrated or disappointed, Jack said he was inspired. It suggested an opening. He resolved that this tool was exactly what the Chinese economy, then rapidly expanding under a government reform program, needed to accelerate its progress and make its products better known beyond its borders.

He had a sense of the potential of the internet for opening markets for China's burgeoning classes of small and medium-sized enterprises (SMEs). When he returned home, he created his first venture, the China Pages, a mere listing service for Chinese manufacturers to display their specialties for business customers. Styled after the thick Yellow Pages phone book, Jack's venture aimed to be an online directory listing the companies in China interested in expanding their markets, domestically and overseas.

It was a sound idea, prescient, but so few used the internet in China— or even knew what it was—that the venture struggled. In 1995 less than 0.1 percent of China's population was online. Strapped for cash, Jack eventually signed a joint venture deal with state-backed Hangzhou Telecom to sustain the business, but he soon found that he had lost control over decision-making. By the fall of 1997, drained by the challenges, he gave up his stake and walked away.

Less than two years later, Jack was ready to try again. He had secured a ministry job in Beijing focused on developing e-commerce but found the layers of bureaucracy stifling, so he again took a look at establishing his own company. Convinced that private enterprise would lead China's e-commerce revolution, he quit his government position and created a new company that helped small businesses connect with the world.

In February 1999, he assembled seventeen close friends and former colleagues and students inside his Lakeside Gardens apartment in Hangzhou and launched Alibaba.com. Jack was never one to set modest goals. "Our competitors are not in China," Jack said at the company's first all-staff meeting. "[They are] in America's Silicon Valley."

China then was still a primarily agricultural society with the majority of people living in the countryside, working on the land. In 1999, the average per capita income was about $800, and there were only 8.8 million internet users in the entire country. I remember my first night in Hangzhou, riding a taxi to my company apartment. The city had at one time been known for its scenery but the years had taken a toll. That first winter in Hangzhou tested my spirits and endurance. The numbing cold easily overpowered the scattered space heaters in our office, and many of us wore gloves at our desks even as we typed.

Beyond the physical discomforts, we confronted the reality that the world Jack imagined was a far cry from the one that we had to work with in the China of 1999. There was no retail e-commerce, no online payments system, and no private logistics services to speak of. But it wasn't only the internet infrastructure that was primitive; many aspects of the commercial infrastructure were also underdeveloped. Even shopping malls and credit cards, basic building blocks of commerce in the United States, hardly existed. In the twenty years after the first Chinese credit card was introduced, only thirty-three million cards had been issued—a fraction of 1 percent of the population. This paled in comparison to the United States around the same time, where more than half of Americans were credit card holders.

Living in Hangzhou, I found that very few vendors would accept my international credit card, and I was never able to locate a working ATM. The closest internationally linked ATM was in Shanghai, a three-hour train ride away. When it came to consumer goods, even the most commonplace items in America were nearly impossible to find. After wearing through my sneakers, I had to check four different stores just to locate a pair of decent Nikes. Tracking down some Quaker Oats oatmeal, a necessity for any good Californian on the road, proved even harder.

I sometimes had to remind myself that I was working at a technology company. The flashy new breakthroughs from Silicon Valley, the excitement and exuberance of that era, seemed to occupy a different universe from the one I inhabited. At the time, the indisputable darlings of the Chinese internet sector were the country's homegrown search portals— Sina, Sohu, and NetEase—all of them located in China's first-tier cities like Beijing or Shanghai.

Alibaba was different from conception. Instead of repurposing a Western model for the Chinese market, its goal was to create a new type of online wholesale marketplace from scratch, aimed at letting Chinese SMEs connect with one another and with new customers. This business required an understanding of how to engage and assist a class of older, more traditional business owners, an entirely different demographic from young, urban netizens. Jack intuitively understood the constraints on smaller enterprises but also China's economic potential and the unique ability of the internet to overcome the barriers.

"It wouldn't matter even if I fail," he once said. "At least I've introduced the concept to others. If I don't make it, someone else will."

It's the Heart Where Wisdom Comes From

Jack was quite open in stating that his vision came not from market analysis, product testing, or focus groups. It came from his heart. He understood the vast potential of the online world, but he also believed that it was just a tool, a means of satisfying human needs, fulfilling human dreams, assisting communities and human progress. He was focused on personal growth and had little fascination with the electrons bouncing around to produce those digital ecosystems. His philosophy permeated the Alibaba culture.

This was driven home during an unusual dialogue that Jack had with Elon Musk of Tesla at a public conference in Shanghai in 2019. These two giants of the digital age discussed and jousted over the importance of various aspects of new technologies, on subjects ranging from artificial intelligence (AI) and space travel to the changes needed in education systems. Musk expressed deep misgivings about the rapid advance of technology and described AI as a potential threat to humanity, along with video games. He worried, he said, that our digital servants could easily become a menace.

"I hope they're nice," he said, when discussing what AI bots might do.

Jack rejected the warnings and refused even to engage with Musk on the technical details of how these tools might evolve or turn into threats.

"Computers only have chips, but man has a heart," he said. "It's the heart where wisdom comes from."

Calling himself an optimist, an adventurer in *inner* space, Jack added, "People will use technology to understand ourselves better, rather than the outside world."

These comments underscored Jack's entrepreneurial genius: it is intuitive, based on his natural feel and deep empathy for human emotions and needs. And luck. It is useful to know that Jack's favorite movie is *Forrest Gump,* and he sometimes describes himself as a Gump-like figure, someone who chanced into the great digital trends of this era but had the intuition to see the exceptional opportunities they opened up and the confidence to seize those opportunities.

Again, China's remarkable progress and Alibaba's growth went hand in hand. Today, China is the first country in the world where e-commerce has surpassed traditional offline retail spending, constituting 52 percent of all retail transactions with over one billion netizens, 99.7 percent of whom are mobile users. Almost all of them use their smartphones to make and receive payments. In 2021, China's e-commerce sales reached over $2.4 trillion, while in 2020 the State Post Bureau reported more than 60 billion express deliveries across the country. Package volume handled by the US Postal Service the same year was 570 million.

While the transformations of China's digital economy over the last twenty years have been nothing short of impressive, to gain a fuller perspective of how this happened it helps to understand which societal factors made this possible both for Alibaba and the country as a whole. In the next part of this chapter I break down the key elements—domestic and international, public and private—that have powered the development of the country's vibrant economy and technology ecosystem.

The Building Blocks of Growth

Key Components of China's Digital Economy

Governance Layer	Data Governance, Platform Governance, Collaboration Governance						
Application Layer: Digital Economy	Education	Digital Agriculture	AI	IoT	Digital Media	Digital Sports	Mobility
	Health	Digital Entertainment	Digital Retail	Digital Manufacturing	Transportation	Shared Economy	Credit
	Rural Taobao	Cross-border eCommerce	Local Services	Digital Financial Products	Smart Logistics	Smart City	Digital Government
Data Layer	Data Property & Assets, Data Flow & Sharing, Data Protection						
Foundation Layer: Infrastructure of Digital Economy	eCommerce Platform		Digital Finance Platform		Smart Logistics Platform		
	Cloud Computing and Big Data Platform						
External Layer: Societal Environment	Integrated Market	Education	Government & Policy	Entrepreneurship		Finance Market	
	Physical Infrastructure			Internet Infrastructure			
	Social Stability						
	Population, Market Size						

(Domestic Market Trends / Global Market Trends)

China's digital economy and the components upon which it is built.

Beginning in 1978, China embarked on a decades-long trajectory of economic liberalization, widely known as "reform and opening" (*gaige kaifang*). Some industries that had been subject to rigid state planning were, little by little, opened to market forces, and privately owned companies were permitted to take root. As the country's economy took off, the private sector expanded, spreading new levels of prosperity. The economy doubled in size every eight years. By 2010, China had overtaken Japan to become the world's second-largest economy. Similarly, a Brookings report stated that since its reforms began in 1980, China has lifted close to eight hundred million people out of absolute poverty, a remarkable achievement by any measure.

A range of societal factors had worked together to help fuel this growth, many of them a tribute to the government's far-sighted policies. Looking specifically at the external layer in the figure on page 29 we can go through the individual factors, listed from the bottom up.

Population

With 1.4 billion people, China's population eclipses the combined population of all fifty-four countries that comprise the African continent. The country's working-age population of fifteen- to sixty-four-year-olds surpassed one billion in 2014—a vital ingredient for an industrializing economy. Companies from all over the world have been attracted to build factories to access the huge pool of inexpensive labor. At the same time, China's rapidly expanding middle class has also given China an attractive consumer market, now the largest in the world.

Social Stability

After decades of upheaval during the Mao era, the Chinese leaders and administrations that followed emphasized social stability as a critical objective, and that supported growth as well as a big influx of human and financial capital. The generally stable environment gave individuals the assurance they needed to make long-term commitments.

Over the past few decades, entrepreneurs were able to build their companies with little disruption, while foreign investors had the confidence to bring in capital. What had been a brain drain began to reverse. As these inflows of talent and investment grew, they increased the speed of this remarkable forty-year run of development. In recent years, China has ranked in the top ten on Gallup's survey of the world's safest countries, just behind top-ranked Denmark.

Notably, some would argue that the country's latest regulatory tightening around big tech and stringent zero-COVID strategies have caused short-term social disruptions. The government's long-term top-priority goal remains, however, the same—to achieve social stability.

Physical Infrastructure

Large infrastructure projects brought clean water and electricity to previously underserved areas, enabling huge leaps in sanitation, health, and economic activity. China's installed power-generating capacity grew at an average annual rate of 10.6 percent through the second half of the twentieth century, or roughly doubling every seven years. As market reforms set in and the country's electric sector privatized, electricity production skyrocketed. Massive highway and rail construction projects provided critical access to cities and rural areas, enabling the movement of people and goods cheaply and efficiently, uniting the country into what is essentially a single, huge market.

To take one example, thanks to the extensive roads and network of bridges, someone living today in rural Gansu, a northern province known for its vast mountain ranges, can order a mobile phone to be delivered from the coastal tech hub of Shenzhen for a shipping cost of 15 renminbi (RMB)—less than $2.50. The package will travel 1,500 miles and arrive within three days. A similar package shipped using the UPS three-day service from Boston to Reno, covering roughly the same distance, will cost more than ten times that amount.

Internet Infrastructure

Since the 1980s, the Chinese government and telecom firms have been investing heavily in weaving a far-reaching network of connectivity. Mobile internet today is now accessible in more than 90 percent of the country, while the average fee for a month of unlimited 4G data is less than half the price of a weekly pass on the New York City subway. In fact in the global league table on mobile data costs compiled annually in 2021 across six thousand mobile plans in 230 countries, the average cost for one gigabyte in China averaged fifty cents and ranked as the world's seventeenth cheapest, compared to the United States at $3.33, which ranks at 154. According to the Ministry of Industry and Information Technology, more than five million 4G base stations have been built across the country—almost half the world's total.

Even in a province like Guizhou, among China's poorest in GDP per capita, every single one of its ten-thousand-plus villages have achieved 4G coverage. And the global pandemic has not slowed down China's deployment of 5G capabilities. In fact, by the end of 2021, powered by the 1.15 million 5G base stations set up nationally, of the country's total 1.62 billion mobile phone users, 497 million were connected to 5G terminals, constituting more than 70 percent of the global 5G subscriber base. The Chinese government expects that by 2025 its 5G network will cover all cities and towns and most villages.

Mobile Everywhere

From payments to shopping, livestreaming to delivery, Chinese consumers today live in a society wholly remade over the past several decades. The catalyst that has propelled much of this forward is a device smaller than a checkbook, which we can no longer imagine leaving home without: our mobile phones.

From the onset, mobile penetration in China has grown at a breakneck pace. In 2000, shortly after I first joined Alibaba, only 10 percent of the population had mobile phones. In 2013, the number of Chinese mobile users eclipsed those in the United States, a major milestone. By the following year, smartphone users were approaching half a billion people. More than 98 percent of the one billion Chinese internet users today are zipping through the web on mobile devices.

The online services that have grown from China's massive mobile adoption form a crucial component in the country's economic restructuring. The devices have helped to accelerate the transition from an export-driven economy to one driven by services and domestic consumption.

In the United States, many trailblazing internet firms developed technologies to create entirely new applications and new industries, such as online advertising, online auctions, and social networking. But as Professor Ming Zeng, Alibaba Group's former chief strategy officer, has noted, the leading Chinese internet firms generally applied new technologies to addressing gaps in *existing* demands.

In retail shopping, for example, the American consumerism boom really began with brick-and-mortar stores, giving rise to retail outlets on street corners and sprawling strip malls and shopping malls in the suburbs. China lagged in retail store construction. Though the United States has less than a quarter of China's population, it has thirty times as many shopping centers; the average American enjoys 3.3 times as much retail floor space as the average Chinese. In China, meanwhile, the physical shopping experience left much to be desired: store were dingy with sparse shelves, too often stocked with a poor selection of products.

As a result, Chinese consumers did not have to break personal shopping habits when they were offered the opportunity to shop online. They were happy to e-shop rather than hunt down a store and deal with a poor consumer experience. As Zeng writes: "China provided fertile ground for [a new] model to unfold."

Of course, smartphones have accelerated this transition to online shopping. From a business perspective, this has enlarged the range of opportunities for retailers. Touch points have increased manifold thanks to geolocation and push notifications, which can alert consumers to promotions and discounts whenever they come within proximity of a store. Opt-in, app-based loyalty programs funnel up valuable data on consumer needs and preferences, allowing e-retailers to provide more streamlined, tailored offers and create a better inside- and outside-of-store shopping experience.

The development of such capabilities has arisen from the integration of four vital business pillars: e-commerce, digital payments, smart logistics, and cloud computing. The ecosystem they have created serves as the basis for many of the new applications emerging today in retail, banking, manufacturing, health care, government services, and other sectors.

Integrated Market

The expansion of reliable physical and digital infrastructure across a more stable, centralized society has consolidated the benefits of a large, integrated market, with few legal or administrative barriers to access from one region to another. By contrast, India, which also boasts a billion-plus population, is divided into twenty-nine states that, in addition to being culturally distinct and using many different languages, have erected a patchwork of tax and commercial regulations, which impede operations across state boundaries.

Education

As China's digital infrastructure started expanding in the late 1990s, so did demand for skilled workers, technicians, and managers. The number of students admitted to colleges increased from 1.15 million in 1980 to 26 million in 2015. And in 2016, nearly 5 million Chinese students graduated with STEM degrees, twice the number in India, the second-highest country, and eight times as many as in the United States. In addition, more resources have been invested in training students to succeed in the digital economy. One Chinese university began offering e-commerce as a major in 2000, and by 2020, students in 563 universities around the country could pursue e-commerce-related degrees.

In one instance, Jia Shaohua, an educator at the state-run Yiwu Industrial and Commercial College in Zhejiang, ignited a mini-revolution when he began offering a radical new e-business entrepreneurship course in 2009. Instead of teaching the traditional requisites, espousing theory and pushing rote memory, he emphasized straightforward, practical lessons on how to start a business. Rather than teach students how to find jobs through vocational training, he endeavored to teach them how to *create* their own jobs for themselves and others. Word of his "enterprise school" spread, with hundreds of students soon signing up. Within a couple of years, many of them had already started their own online businesses.

Professor Jia's graduates produced, marketed, and sold everything from knitwear to cosmetics, generating tens of millions in RMB sales and creating more than a thousand jobs. His novel training model was upheld as a model for preparing college graduates all around the country.

Government and Policy

The dynamism of China's public-private partnership is one of the key elements that has enabled the digital economy's growth over the past twenty years. Although many would argue conditions have recently changed dramatically for many internet companies and are not as favorable, the Chinese government's initial willingness to grant entrepreneurs a relatively free hand to experiment and its early economic reform programs promoted the rapid growth of China's digital economy. In many cases, authorities moved to impose new regulations or restraints only after a deliberate delay. In that way, China's internet giants enjoyed relative freedom for years before the government took any concrete regulatory actions. With Alipay, for example, Alibaba's pioneering payments service, eleven years passed after the introduction of online money transfers before regulators set a cap on the value of the transfers.

Generally, the Chinese government offered guidance and incentives through its policy announcements instead of imposing onerous restrictions. The "Made in China 2025" and "Internet Plus" plans, both launched in 2015, presented clear objectives for the country's industrial and digital development. The Five-Year Plan for National

In the early years of mobile payments, government policy left space for innovation

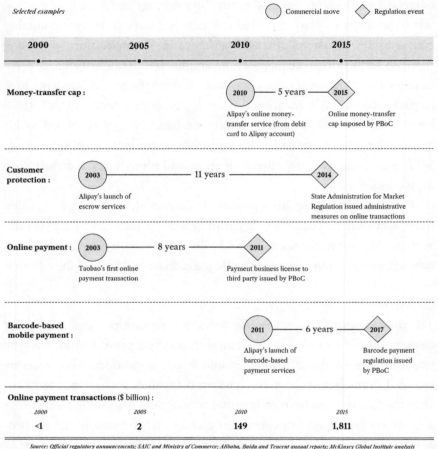

Selected examples   ◯ Commercial move   ◇ Regulation event

2000   **2005**   **2010**   **2015**

Money-transfer cap :

2010 — 5 years — 2015

Alipay's online money-transfer service (from debit card to Alipay account)

Online money-transfer cap imposed by PBoC

Customer protection :

2003 — 11 years — 2014

Alipay's launch of escrow services

State Administration for Market Regulation issued administrative measures on online transactions

Online payment :

2003 — 8 years — 2011

Taobao's first online payment transaction

Payment business license to third party issued by PBoC

Barcode-based mobile payment :

2011 — 6 years — 2017

Alipay's launch of barcode-based payment services

Barcode payment regulation issued by PBoC

Online payment transactions ($ billion) :

2000	2005	2010	2015
<1	2	149	1,811

Source: *Official regulatory announcements; SAIC and Ministry of Commerce; Alibaba, Baidu and Tencent annual reports; McKinsey Global Institute analysis*

With many introductions of new technology, Chinese regulators waited for a period of time before stepping in.

Informatization, introduced the following year, pledged greater re-sources for technologies from 5G networks to cloud computing and the internet of things—meaning the physical objects (sensors, software, technologies, and devices) that exchange data over the internet. And in 2017 China's State Council released the country's strategy for develop-ing AI, entitled "New Generation Artificial Intelligence Development Plan." These were road maps signaling the government's economic pri-orities and favored investments. It left it to entrepreneurs to determine how those priorities would be realized. In essence, the Chinese system has permitted market forces to spur competition. The best survive and thrive. Only then do regulators step in and determine how the market sector will be managed moving forward.

In August 2018, the Chinese Congress passed the world's first e-commerce law, which included the regulation of contracts, logistics, operations, and payments. But this followed five years of measured deliberation, four official reviews, and three public consultations. More recently, the government has granted tax breaks and investment incentives to venture capitalists, encouraged the flow of talent to the best companies, and even allocated entire city districts for the express purpose of supporting innovation.

The environment has shifted in the last two years with the government tightening policies with stricter enforcement. And although government overtures in June 2022 signaled an easing in some of this pressure, some uncertainty remains as to how the situation will unfold. It has made clear its concerns over data privacy, anticompetitive practices, and the sheer power amassed by the largest technology companies.

Nonetheless, for the first two decades of China's internet sector, this broad *late*-touch approach gave innovators the space to experiment and tinker, testing out various strategies for commercializing their products to gain critical mass. This open environment had benefited Alibaba, enabling it to become both an innovator and, itself, a benefactor. Jack Ma often recognized that the company could not have achieved its far-reaching impact without the favorable climate created by the public sector.

Entrepreneurship

At the heart of it all has been a generation of Chinese pursuing entrepreneurial dreams that were absent for many years. In 1949, the Chinese Communist Party came to power and replaced the nascent market economy with a Soviet-inspired socialist model. Production and commerce were subject to central planning. Private businesses were tightly controlled by the state until Deng Xiaoping initiated the reform and opening up era. Quotas fell away, new incentives were introduced, and a spirit of ambition and innovation reawakened. China's entrepreneurial sector sprouted more than six million registered private businesses by 2014, accounting for more than two-thirds of the country's GDP. In 2016, China recorded 1.3 million patent applications—more than the United States, Japan, South Korea, and the European Patent Office combined.

One of the engines of this national revival has been the coastal province of Zhejiang (home of Alibaba), which was the home of a vibrant entrepreneurial culture for years. In fact, today Zhejiang is one of the most prosperous and economically diversified provinces in China with a GDP in 2021 of $1.14 trillion. It is home to over 2.22 million SMEs and hosts ten of China's top one hundred private-listed companies, and is one of the country's most vibrant high-tech start-up communities.

Finance and Investment

China's venture capital (VC) sector—nonexistent when I first arrived in China and barely a blip even a decade ago—has grown into a formidable force. By the second quarter in 2017, Chinese VCs actually surpassed North American funding, receiving 47 percent of the world's VC funding. Thanks to VC investment, China has filed more blockchain patents than any other country and ranks in the global top three for investing in areas like AI, autonomous vehicles, and robotics. Amid the coronavirus pandemic in 2020, China surpassed the United States to become the world's top destination for foreign direct investment.

Alibaba was greatly assisted by all of these factors, and Jack—like Forrest Gump, with his uncanny knack for showing up at critical historical moments—was both a savvy entrepreneur and evangelist who brought powerful insights into how to deploy and market his platform. While some might say Jack's enthusiasm appeared to border on naïveté, it was actually based on his intuitive understanding that he was riding a wave of economic expansion generated by the Chinese government's deep commitment to integrating with global markets.

It's amusing to recall the challenge we initially faced in getting SMEs interested in this strange new digital network when we were trying to rev up our marketplace. We saw this vividly when, in 2000, we organized Alibaba's inaugural offline meetup at the Hongqiao Hotel in Shanghai.

Jack introduced our company in a banquet hall with about two hundred attendees. The audience was comprised of mostly small business owners, primarily manufacturers and exporters from the Shanghai and Zhejiang areas. They were men of middle age, dressed in thin polo shirts and loose jackets—not digital natives. Many had heard of the internet before but knew little about how it worked. Jack explained that it was Alibaba's mission to link Chinese suppliers with the rest of the world, and he ended his talk by encouraging everyone to seek out Alibaba representatives to learn more. We had prepared by giving our staff blue-and-tan baseball caps with the Alibaba logo. The attendees clustered around my colleagues and filled the air with questions.

"How can Alibaba help me?

"Who are your buyers?"

"Why should we trust you?"

"How much does it cost?"

We described Alibaba as a trade show on the internet. Instead of paying to register at these crowded, often chaotic conventions, and then lugging bulky suitcases stuffed with product samples to city after city, Alibaba offered a tidy virtual display to showcase everything they had to offer. And instead of having to stand beside a physical booth for weeks, simply hoping the right clients would walk by, Alibaba enabled managers to present their listings to interested buyers twenty-four hours a day, 365 days a year—without ever needing to leave their offices.

The curiosity and excitement in the room was electric. We felt like missionaries, proselytizing to these underserved entrepreneurs about this entirely new concept—the *internet*—that would transform their business.

Global Market Trends

One afternoon not long after I joined Alibaba, I found myself standing around Joe Tsai's cubicle with Jack and a few other colleagues from the international team. Jack was brainstorming about appointing a group of business advisers, prominent figures who could offer guidance and help raise the company's public stature in China and outside. He suddenly broke out in a grin, which we had all come to recognize as a sign that an idea was taking shape.

"I know! Bill Gates can be our adviser."

We all laughed. Perhaps one day, if everything went right for us, we might attract such a leading technology icon, but we all recognized that that wasn't likely to happen anytime soon.

"OK, fine, we'll save Bill Gates for later," Jack said, feigning disappointment. He got serious again. "Let's think beyond tech. We're trying to promote international trade *through* tech, so what are the respected organizations out there involved with international trade?"

Someone piped up with the obvious answer: the World Trade Organization. Founded in 1995, the WTO is a forum for rule making and overseeing the operations of a free and fair global trading system. Its 164 members represent 98 percent of the world's trading activity. China had sought to be included as one of the organization's founding members, but other member countries insisted that first the Chinese economy had to further liberalize and open up to investment. Following the 1997 Asian financial crisis, the Chinese government instituted many of these reforms and was progressing toward admission.

Jack nodded slowly, considering the possibility. "Yes, that's a perfect fit. If we could get the head of the WTO, that would be amazing." We agreed, though to us it seemed like just as much of a pipe dream as Bill Gates.

Then Joe Tsai spoke up. "Actually, I think there's a possibility there." Joe explained that through his contacts we could get in touch with Peter Sutherland, the WTO's founding director-general, who had since retired from the organization. "It's definitely still a long shot," he said, "but it's possible we can at least make the inquiry."

It clicked. Within a few months, we had contacted him and reached an agreement. In April 2000, we announced that Sutherland had become Alibaba.com's board adviser. The appointment helped elevate the company's brand recognition and highlighted our commitment to forging deeper commercial bonds with the world. It lent us greater credibility.

Jack was right in his assessment: Alibaba never started as a company boasting the most cutting-edge technology. Rather, its fortunes were inextricably bound to the fortunes of China's growing tide of SMEs and their ability to access new markets as players in the global trading system. That inclusivity was key for Jack and our working model. We would continually build on this objective by expanding Alibaba's operations, step-by-step, into the new digital economy's four core pillars: e-commerce, digital data, logistics, and cloud computing. We would create an ecosystem of interacting applications to assist SMEs in mastering every aspect of the digital economy.

From a historical perspective, Alibaba had dropped in on this wave right before its crest. It was an example of Jack's Gump-like good fortune, being at the right place at the right time—a trait that he has conceded. In fact, in numerous interviews, Jack has pointed to that character as an inspiration to him about the importance of persistence and a belief in one's self, no matter the odds. Jack's vision of creating this B2B marketplace helped bring an enormous number of Chinese suppliers and manufacturers online, giving them access to domestic and global markets. China's entry into the WTO was formally approved in December 2001, broadening this market access.

Domestic Market Trends

Beyond China's borders, the world was finally waking to the Chinese internet sector's potential. A widely circulated analyst report in 2004 laid out the proposition. "Investors still underestimate the impact the Internet will have in changing business process and consumer behavior on a global basis—and we believe that China is emerging as a market that helps prove this point," the report said. Meg Whitman, then CEO of eBay, tech's shimmering success story at that time, predicted: "Whoever wins China, will win the world."

Jack certainly had believed that for years and was continually planning to ensure that Alibaba came out on top. In early 2003, he had pulled together six trusted lieutenants into his Hangzhou office and sworn them all to secrecy. Realizing that it was only a matter of time before eBay attempted a full-scale campaign in the Chinese market, he assigned the group to a confidential project, creating a website to compete directly with them. On May 10—a date that would become our annual Ali Day celebration—Alibaba launched its own online consumer marketplace, Taobao, meaning literally "digging for treasure."

Taobao proved to be a big hit, making immediate inroads into eBay's market share. Within two years, it had already surpassed eBay in both web traffic and product listings. The American tech giant had struggled in adapting to Chinese culture and consumer preferences since many of its decisions were being made from its headquarters in Silicon Valley. But Taobao had made customer satisfaction a focus from the start, localizing features, including elements of webpage design and a well-publicized pledge not to charge any fees for three years to encourage customer loyalty.

For instance, we found through our research that Chinese netizens often like web pages that are livelier and busier, with more content flashing through the screen. It is reminiscent of the kind of Asian street markets that many Chinese are familiar with, filled with stimulation, colors, people, and merchants. So, based on our research, we designed livelier pages for consumers to peruse. In addition, the researchers studying how consumers scanned the pages found that they frequently would read from the top left to the right, then scan downward, so materials were lined up to follow that eye pattern.

And we deployed an Alibaba instant messenger tool called Aliwangwang. This gave Taobao consumers the ability to interact with merchants quickly before making purchases, getting product information and even bargaining, much like a real street market. Since eBay's model depended on taking a sales commission, there was not a fast, direct communications link between buyer and seller prior to the purchase, so Chinese consumers did not have the same degree of interaction. These were important features of our platform.

Alibaba's rise during those years also reflected domestic macro forces and how the Chinese economy was developing. As incomes rose and the market for goods and services expanded, many Chinese embraced a more comfortable and materialistic lifestyle. Gone was the caution and economic fear of previous decades. Taobao's easily navigable website, quality assurance tools, and attentive shopping perks helped shepherd in a new class of confident and informed consumers. China was no longer only making things for the rest of the world to consume; it was buying them, too.

After three decades of a supercharged economic expansion reliant on heavy industrialization, manufacturing, and exports, China was beginning to exhaust its old growth model. With some sectors saddled with overcapacity, the government announced a reform program to shift the economy toward new growth through services and domestic consumption. Jack's digital platforms played an important role in achieving this economic restructuring.

I saw firsthand the reality that, while some big international companies like eBay understood in theory how important the Chinese market could become, China still seemed, for many, exotic and faraway. Even younger businesspeople, who tended to see themselves as daring and forward looking, overlooked it.

Parallel Worlds

When I left Alibaba for the first time in 2001 to attend Wharton, I was proud that I had wandered far from the accepted pre-MBA path and had worked at an unusual, little-known Chinese start-up. I had been adventurous and had learned a lot, but many of my classmates either had worked at big Wall Street banks or wanted to, and the internet industry had lost much of its sheen with the bursting of the so-called dot-com bubble a couple of years earlier. I was a little bit of an outlier.

One day I heard from a former colleague that Jack was coming to the East Coast to visit investors, so I reached out and invited him to give a talk at Wharton on how the internet was impacting small businesses. Jack loves speaking with young people, so he readily accepted. It was the fall of 2002.

Typically, such events attracted sizable crowds, since the students are often quite eager to introduce themselves to these business celebrities and the job prospects they represent. I was an officer of Wharton's Asia Club, so I recruited its members to attend Jack's talk and circulate word of the opportunity to hear this interesting entrepreneur. I sent emails to other student groups, taped orange flyers to walls and lampposts around campus, and booked a nice-sized auditorium. But by the time the day arrived and we approached the scheduled 5:30 start, it was clear that my efforts had been in vain. To my embarrassment, barely a dozen people had shown up. Many of them, I noted grimly, were the other officers of the Asia Club.

If Jack was put off by the low turnout, he didn't show it. Instead, in classic Jack fashion, he opened with a jab at the value of MBAs—a recurring theme in our conversations over the years—before launching into a presentation on how tech firms like Alibaba were helping SMEs become a part of the world's business landscape and transforming it in the process. This was not a talk about techniques for managing dozens or hundreds of people, for how to apply performance metrics or improve returns—typical business school fare—it was a vision of a new approach to business, delivered with Jack's infectious energy.

After his talk, a few of us squeezed into a rented car, drove into downtown Philadelphia, and found a small Chinese restaurant, where we slurped our meals from a steaming hotpot and talked about the world as Jack saw it. He was upbeat, as usual, and I thoroughly enjoyed reconnecting with someone I regarded as a mentor. Perhaps most important, I reconnected with an appreciation for how conservative business thinkers can be and how elusive opportunity can be.

3

Alibaba's Ecosystem

The Everything Platform

A chattering alarm clock jolted me from a deep slumber. It was 5 a.m., barely two hours after I had finally gone to bed. I shook myself awake and rose to greet the morning of September 19, 2014—the day of Alibaba's much-anticipated American IPO.

By 6 a.m. Jack and I were cruising down FDR Drive toward the New York Stock Exchange and into the history books. The Alibaba Group's listing was widely expected to be the largest public offering by the NYSE in its more than two-hundred-year history. Jack sat quietly, switching his attention back and forth between his phone and the flickering lights of New York. It was still dark, the sun just starting to rise. The autumn air, blowing through the windows, was crisp.

Given the momentous stakes, I couldn't believe how calm Jack appeared to be. For me, memories of our rambunctious, almost playful past and the weightiness of the present moment were colliding. Despite all the work everyone at Alibaba had put into getting us to this moment, the reality was still sinking in.

In June, we started preparing for our investor road show, an intense flurry of meetings around the world. Stateside, we met with most of the largest investment firms—T. Rowe Price in Baltimore, BlackRock in New York, and Fidelity in Boston. In Asia, we visited the investment offices of regional sovereign funds and local family offices in Hong Kong and Singapore. No matter where we traveled, it seemed, there were headlines about Alibaba. The expectations were enormous.

As the IPO date drew closer, we'd broken into two groups. One returned to the United States to continue the investor meetings. I headed for the Middle East with Jack, bouncing from country to country, concluding the trip with meetings in Dubai. On our flight back to New York, Jack played cards with friends, while I tried, and failed, to catch some rest. Somehow, we had managed to fly around the world twice in two weeks in an effort to explain to investors what Alibaba was, and now we were back.

On the evening of September 18, we all gathered for a final meeting of the road show teams, our bankers and lawyers, even some of the company's earliest investors such as Masayoshi Son, the founder of SoftBank, to approve the final pricing of the IPO. As we raised glasses and toasted the milestone and the opening of Alibaba's exciting next chapter, I stood there witnessing what was an amazing feat of preparation, coordination, and execution across so many parts of the world by this amazingly dedicated team. I was awed by the size of the effort. At the same time, as I sat in that Manhattan office tower, I also wondered if my colleagues in Hangzhou realized the full impact of their work.

Our car finally pulled up onto Broad Street right in front of the New York Stock Exchange, where a pool of reporters and photographers were waiting. I felt a jolt of pride at the sight of the row of classical columns on the NYSE's famous facade, festooned with massive orange-and-white Alibaba banners. All for us. It was surreal.

The morning soon spun into a whirlwind. I tried to assist Jack through a chaotic swirl of meet and greets, photo shoots, interviews, and speeches. At times I felt overwhelmed, but Jack, calm and composed, was gliding along with a smile under the spotlight.

As ever, Jack was more of a philosopher in his comments than a typical data-driven CEO. For him, the theme of the day was his heart, not his head. In one interview, with CNBC, Jack was hit with a barrage of financial questions—about valuation, share price, and GMV (gross merchandise value). Jack being Jack, what he wanted to stress, as he put it, was *trust*. He used the word eight times—not technobabble.

"Today what we got is not money," he said. "What we got is the trust from the people."

As the flood of appraisals of the size of the offering and the rise of the stock price continued over the following months, I came to feel, too, that much of the attention was missing the point. It was important to assess how much value was going to shareholders, but Alibaba's *true* goals, as always, went beyond income statements and balance sheets. Alibaba had built not just a great company but a powerful new model—a digital ecosystem. It was expanding the boundaries of technology and exerting a huge economic and social impact on China and other countries, unmistakably advancing the common good.

What Jack was trying to tell the world was that this was what e-commerce could produce, not just wealth but a new type of inclusive prosperity and positive interconnectedness. Alibaba was redefining, in a subtle way, what it means to be a "successful" company in the twenty-first century. I hoped the rest of the world would recognize this aspect of the Alibaba innovation.

Jack Ma's North Star

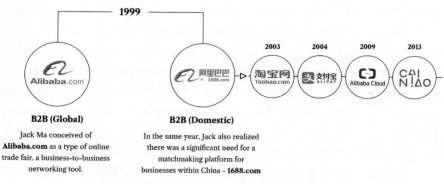

The evolution of the Alibaba ecosystem.

It's easy to see Alibaba's journey as a seemingly preordained march from one triumph to the next. But there were many bumps along the way. Most notably, there was our near bankruptcy in 2002. As Alibaba was running perilously short of cash, we had to be rescued at the eleventh hour by Japanese investors.

Later, we almost became a victim of our success. The volume of online transactions was so large and growing so rapidly that we realized, almost too late, the costs of outsourcing our computer server needs for managing all the transactions was starting to devour our sales revenues. Before efficient delivery services were built that were large enough to handle the volume of packages being shipped during major sales promotions, Alibaba found that some packages took months to get delivered. New ideas and market fads constantly tempted us to swerve in new directions, and we had to be careful when evaluating opportunities.

Jack's approach was to strip everything down to the basics; not the basics of cash flow, earnings per share, return on investment, or the rest of the Wall Street metrics, but the basics of human need, of enriching human progress. Need creates opportunity—that is Jack's North Star. The rest of us had to figure out the details. The greatest need he saw was in expanding opportunities for China's huge number of industrious SMEs and entrepreneurs.

When he started Alibaba, as clever and hardworking as many of those entrepreneurs were, few had the capability or the know-how to access either the broader Chinese national market or global markets. In 1995, his first company, China Pages, had faltered because the internet was too unfamiliar and many businesses were hesitant to try something that seemed so exotic. And after partnering with the local government to try and salvage the effort, Jack had faced a different problem as the venture was forced to shift its focus to serve not energetic entrepreneurs but China's bulky legacy firms, many of them state owned. He was losing sight of his North Star, and so he chose to leave that first company behind.

Jack then went to work for Infoshare Technology in Beijing, the e-commerce center of the Ministry of Foreign Trade and Economic Cooperation (MOFTEC). But after just a year in Beijing in that role, Jack decided that the best way to pursue his entrepreneurial passions was working outside of government back in Hangzhou.

By the time Jack had regrouped and launched Alibaba, in 1999, technology firms had rocketed to center stage and were the darlings of the US markets. A wave of ambitious Chinese start-ups were vying to emulate that dot-com model. But while China's early tech successes like Sina and Sohu focused on web-based search, essentially emulating Yahoo!, Jack and his cofounders built a very different business.

China had achieved success as a manufacturing and export economy, and negotiations were underway for China's entry into the WTO, both a practical achievement as well as a badge of honor, a sign that this once impoverished nation had arrived on the world economic stage as a real player. China's vast potential was becoming clear as the world's factory floor. What Jack understood was that Chinese SMEs, not just giant manufacturers, had the skills and capacity to meet quality standards and produce goods for the global markets. They just lacked an affordable, convenient way to find buyers overseas.

That was why he initially conceived of Alibaba.com as a type of online trade fair, a business-to-business networking tool.

The Leap to Long-Lasting Fortune

It initially operated as a simple bulletin board system, or BBS, for sellers to list basic facts about their factories and products for foreign buyers to browse. That was the design, but within a few months, Jack had another epiphany, that there was a significant need for that matchmaking platform for businesses just within China. So Alibaba launched a second online marketplace in 1999, devoted to helping Chinese wholesalers connect with business buyers within China. That platform would come to be called 1688.com—a homonym for "long-lasting fortune" in Chinese.

Little by little, tens of thousands of Chinese SMEs grasped the benefits of these digital networks, which removed barriers to broader market access. By 2001, China was admitted to the WTO, and Alibaba was on its way to signing up one million members. Alibaba.com fundamentally transformed Chinese firms' understanding and acceptance of the internet as a business tool.

In 2003, Alibaba.com finally broke even after years in the red. But before anyone could breathe a sigh of relief, Jack announced his next ambition in constructing his digital ecosystem—to launch China's leading consumer-to-consumer (C2C) marketplace, called Taobao.

From Bad Stores to a New Type of Shopping Experience

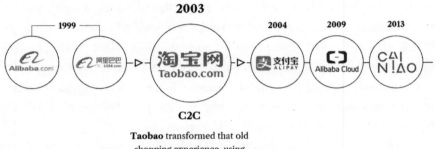

Taobao transformed that old
shopping experience, using
consumer-to-consumer networking
to greatly improve choice and
convenience.

*Taobao's creation opened up a new world of
Chinese retail consumption.*

Before the arrival of the internet, China's retail shopping system was badly underdeveloped. Most Chinese had limited access to the consumer goods they desired, especially outside the major cities. Stores were often dark and poorly stocked and the staff less than helpful. The quality and selection of consumer goods was poor. It was difficult if not impossible to shop for bargains or compare prices.

Taobao transformed that old shopping experience, using consumer-to-consumer networking to greatly improve choice and convenience. With these extensions of the platform, Alibaba helped bring e-commerce into the mainstream for Chinese internet users, netizens, and traditional retailers alike and helped create the opening for Alibaba's next e-commerce ecosystem innovations such as Tmall (business-to-consumer marketplace), Juhuasuan (group buying and daily deals site), and Xianyu or Idle Fish (a secondhand and used goods marketplace).

At the time of its launch, Taobao faced unique conditions and challenges that had to be overcome, yet it pursued its core mission of responding to a need, facilitating digital commerce within a country yearning to catch up with its developed peers and creating a foundation for improvements in the quality of life. Taobao's ambitions were well timed. Against a backdrop of slowing growth in heavy manufacturing and exports, Alibaba's dream of expanding e-commerce aligned with a new government initiative to promote domestic consumption.

The empowerment of SMEs by Taobao and its sister platforms created opportunities that earlier generations could have never imagined. Many who previously toiled in factories or kitchens were now choosing to start their own ventures and—in both big urban areas and smaller towns—become rich. These plucky digital entrepreneurs formed the early waves of a movement that would create a new generation of e-commerce warriors.

Breaking Down the Barriers Between Money and Trust

Payment

Alipay spurred the creation of an entirely new approach to financial services powered by data.

Alipay addressed the lack of trust in the e-commerce market.

When Taobao first launched, getting average Chinese to shop online was like scaling a wall of ice. It was seen not just as difficult but threatening. Hardly anyone was willing to try it.

In the United States, Americans had accepted e-commerce sites like eBay and Amazon naturally. This was due in part to the ease of conducting these transactions using extremely familiar electronic money. Bank-issued credit cards had reached more than half of American households by the mid-1980s; by 2000, more than two-thirds of American families had at least one.

In China, however, less than half a million credit cards circulated in the entire country of a billion-plus people at the end of 2002, barely registering as a percentage of the population. China remained a largely cash economy, and people were simply used to transacting their everyday needs with what were often wrinkled wads of filthy bills.

Modifying habits was not going to be easy. There was almost no trust in anything but face-to-face transactions, with payment in full at the time of sale. Conducting business *online*—especially sending payments or products to a faceless figure at the other end—struck many as an outrageous risk. Even as more Chinese became acquainted with the internet, transacting online was a big ask. The trust just wasn't there.

One big barrier was banks. At the time, the financial industry in China was dominated by large state-owned banks, and government regulations prohibited private firms from settling financial transactions. In the early years, Alibaba.com was not a party to the online transactions: it merely brought the buyers and sellers together, and they handled separately the transactions with letters of credit or other traditional financial instruments.

Producing a more efficient payments system seemed an impossible hurdle until January 2004, when Jack attended the World Economic Forum's annual meeting in Davos, Switzerland. Listening to the panels and lectures that year, he noted, approvingly, that nobody was talking about making money. Not a single person he met seemed to be focused on extending their bottom lines or maximizing profit. They discussed instead new ideas for solving social problems, exciting projects to impact their communities, and their aspirations to change the world.

This delighted Jack. It was all the motivation he needed to take the next big leap in Alibaba's evolution as a high-functioning digital ecosystem precisely because, he believed, it was ordinary Chinese who had the most to gain from a smoother and less expensive means of conducting transactions. Jack called Alibaba's headquarters in Hangzhou. "Do it," he said to his team. "Move forward with Alipay right now."

Concerned about the way China's rudimentary payments procedures were holding back e-commerce growth, a team had already been drawing up plans for an online payments tool that might resolve the issue. Yet the risks of crossing the government's banking policies had held them back from actually launching the new capabilities. Now, inspired by the dialogue at Davos, Jack had greater confidence that it was time to make a move. "If we do not do this, China may miss the opportunity to build its e-commerce," he told his colleagues. "This is something we must do."

In fact, Alibaba's move into an electronic payments function had come as a last resort. After many attempts to negotiate with banks in China's sluggish financial sector, Jack and his lieutenants concluded that breaking through this severe bottleneck required an entirely new system. Many of the transactions on Taobao were valued at little more than five or ten yuan—amounts too small for large banks to care about. Alibaba had to build something from scratch.

It was a colossal challenge but one Jack relished. His first order of business was deciding who should lead the effort, and his impulse was to seek someone unconventional. Jonathan Lu was a former manager at the local Crowne Plaza hotel, who had been hired among Alibaba's first wave of employees, and who served as Alipay's founding president and eventually rose to CEO of Alibaba Group. He excelled at the time leading sales in the Shenzhen region, and Jack called him in for a meeting. Jack later recounted their conversation, proudly.

"Jonathan, what do you know about financial services?"

"Nothing," Jonathan said.

"Good," Jack said. "How about PayPal—can you tell me about them? Have you ever used their service?"

"No," Jonathan replied.

Jack lit up. "Excellent! You're the right guy to lead this new project for Alipay."

Capricious as that process sounds, Jack had a strategy. He knew he did not want anyone with lingering biases or habits from the hidebound banking and financial sector, whose experience might bog him or her down in the industry's traditional ways of doing things. Industry credentials mattered little. This was a new business in a new digital world and required new thinking. Jack needed someone who understood, first, what customers truly needed—an ability Jonathan had demonstrated in his previous roles.

When Alipay launched in December 2004, its basic structure was simple and easy to explain—and trust. Buyers could send their payment to a third party, Alipay, which would only release the money to the seller after the product had been received by the consumer without complaint. The escrow-based system alleviated China's trust bottleneck, assuring customers they would get a prompt refund if their products didn't arrive, while giving merchants confidence that the sale would be consummated and paid for. Though the mechanics of Alipay started out straightforward, they led to a thorough reimagining of the Chinese payment and lending environment for the first time. Another critical dimension of the digital ecosystem was put in place.

Data Leading the Way from Payments to Lending

Alipay started as a simple payments system but quickly began impacting many aspects of everyday life. As it evolved from a desktop tool just for Taobao transactions to a system available for other platforms and then expanded to mobile payments, Alipay gave everyone from local street merchants and craftsmen to workers and pensioners access to a significantly easier means of navigating common transactions, reducing the friction that had long made many of those transactions an ordeal. Alipay began to change those attitudes with remarkable speed because it worked. It institutionalized trust. Jack clearly understood the deeper impact of his innovation.

"You are buying things from somebody you have never seen," he said. "You are giving products to a person you have never met. I want to tell the people that the trust is there. Because it is all about the trust."

In 2011, Alibaba Group spun off Alipay into a separate business that was eventually renamed Ant Financial. (Ant Financial rebranded again in 2020 to its current iteration, Ant Group, offering a broader array of financial and lending services.) The digital ecosystem had streamlined a once cumbersome process, attracting sharply increasing numbers of people into digital commerce.

China's financial services sector was ripe for even broader disruption, and a key tool for remaking the industry was data. Everyone who participates in e-commerce creates a record of his or her transactions and behavior online, like footprints in the sand. The more people that sign up for a platform and the more often they use it, the more data they generate, which can be analyzed to provide a clearer understanding of how consumers think and act and what they want.

As traffic grew, Alipay was able to collect enough user and transaction data to conduct research and find patterns in other areas of the market, particularly for underserved populations. One obvious area of need was credit. At that time, smaller businesses found it difficult if not impossible to borrow money from banks to finance operations and expansion. Jack and the team recognized that they could apply analytical techniques to the volumes of data Alipay was collecting and create a safe business lending platform. Eventually, what became Ant Financial initiated its own lending division.

At that time, most bank loans from the traditional financial sector were only available to large and established firms, which had collateral and could devote sufficient personnel to satisfy the banks' onerous due diligence and reporting requirements. By analyzing user data generated on Alipay, Ant Financial learned it could build credit profiles that were just as accurate as anything a bank could prepare, if not more so. Armed with these innovations, the company has filled an enormous need, extending loans to the 80 percent of SMEs that had been locked out of the traditional banking sector.

The same principle of inclusive finance propelled the creation of another extension of the ecosystem model, Sesame Credit. Just as credit cards never achieved critical mass in China, China never embraced a single, widely accepted credit-rating standard such as FICO or Experian in the United States. By expanding upon the insights drawn from its new lending service, Ant Financial established Sesame Credit in 2015—a credit-scoring service that crunches "a thousand variables across five data sets" to determine an individual's credit score between 350 and 950. For the Chinese market, this was a genuine breakthrough that further broadened access to finance and credit.

Those with high scores on Sesame Credit could enjoy certain perks: renting cars, bicycles, portable power banks, even umbrellas, without having to put down deposits. In a country where only one in three people had a bank account as recently as ten years ago, Sesame Credit built an alternative system that vastly extended and institutionalized the impact of trust. Another piece was falling into place.

The Cloud Computing Experiment

Cloud Computing

"I don't know much about cloud computing and I don't know much about technology. But I truly believe that this will be the future..."
- Jack Ma

Alibaba Cloud established the foundation for a powerful computing resource.

I still remember where I was when I learned about Alibaba Cloud—the cloud computing and network business that would form the crucial third pillar of the Alibaba ecosystem. It was September 2009, and I was on a stage in Hangzhou, making final preparations as an emcee for that wonderful tenth-anniversary celebration. Out of the corner of my eye, I noticed an unfamiliar logo projected onto a large screen behind us. It just said "Alibaba Cloud."

I looked around at my colleagues for a possible explanation, but they were as puzzled as I was. It was a collective head-scratching moment.

We all knew what the cloud was, an innovative means of storing and crunching data run by third parties, but we had heard nothing about Alibaba getting into the business.

As we would soon learn, a team of engineers had begun work on a groundbreaking cloud-based computing service over the previous year. Led by Dr. Wang Jian, formerly with Microsoft, they were inching toward a launch of the new service. Their contribution would help secure Alibaba's future for many more anniversaries to come. In its early days, however, the fledgling division's path was not a smooth one.

In 2009, Alibaba's top management was debating what new initiatives to fund. Some saw mobile devices as the next great frontier. Apple had introduced the first edition of its iPhone a couple years earlier and was enjoying explosive growth. Lots of tech firms were quickly repositioning to navigate the massive changes in consumer behavior this little device was introducing.

A smaller camp warned that Alibaba needed to build out its own data operations in order to support the tremendous growth on its e-commerce platforms. The choice, in other words, was between catching a powerful emerging trend or fortifying the company's unsexy but necessary structural capacities.

During a financial review meeting, a staff member asked Jack how much profit he expected the company to earn over the next few years to help shape the planning process. Taobao, which had become Alibaba's flagship website, had by then eclipsed eBay and asserted itself as China's top online retailer. The question appeared to be a softball, cued up for Jack to deliver a big audience-pleasing forecast.

Instead, he shocked the room. Jack remarked, matter-of-factly, "Alibaba will soon go bankrupt."

Once the initial discomfort subsided, Jack went on to explain that most of the company's revenue was being funneled directly into third-party software and IT providers, which were hosting the domains that supported Alibaba's platforms. This was a massive job, and so the costs were substantial. It was a conundrum of one step forward, two steps back—the more traffic these online marketplaces enjoyed, the more expensive it became for Alibaba to keep things running.

Put simply, we had not thought out our infrastructure needs clearly enough, particularly in a situation where volume was expanding so rapidly. The company's ability to generate online transactions was outpacing its ability to service and sustain them. Developing our own servers to manage this structure in a cost-efficient way was no longer just a good idea—it was existential.

In China and elsewhere, the business world was waking up to the true utility of data, especially big data, the ability to capture and analyze huge volumes of market information, and then applying sophisticated tools such as AI and algorithms to discern patterns, forecast trends, and design new products to meet demand. But, as Alibaba discovered, this can be costly. Jack came to believe that no large e-commerce enterprise would survive long term without developing and owning its own servers and data-processing capabilities.

At the time, most tech firms operated on commercial software supplied by a handful of large established multinationals. Wang Jian set out to unplug Alibaba's platforms from these third-party providers, shedding the reliance on IBM mainframes, Oracle databases, and EMC data centers. His goal came to be known as Alibaba Cloud—also known as Aliyun (*yun* is the Chinese word for "cloud"). The aim was to fundamentally rebuild the company's IT infrastructure, tying it to a network-enabled cloud system that he was developing.

This was seen as a risky leap. Across China, internet companies were buzzing over the promise of cloud computing, yet few had tested what these concepts would look like in practice. Wang acknowledged great uncertainty, but he was firm. "We were confident we could do it," he said.

An Army of Five Thousand Servers Marches into the Future

Not everyone in the company shared his confidence. Though Alibaba was a proven leader in e-commerce, developing enterprise software on our own was far outside the company's core competencies, made even more difficult by the fact that this was a young and still developing field.

But we went ahead, and when Alibaba Cloud's promised benefits did not occur quickly, the heads of other business units were quick to criticize the risky effort.

Alibaba's internal BBS was filled with messages attacking Wang Jian's credibility and his team's lack of progress. Even though Jack had personally committed significant resources and personnel to transitioning Alibaba onto a cloud-based IT system, the pushback continued. Some called for Wang to be fired, while others tried to poach his disgruntled engineers for their own business divisions. Rumors spread that Jack's patience was wearing thin, that Wang's days were numbered, and his cloud project was on the chopping block. Jack finally stepped in. In a public display of reassurance, Jack doubled down on his belief in Wang and demanded an end to the corrosive criticisms.

"Look," he said. "I don't know much about cloud computing, and I don't know much about technology. But I truly believe that *this* will be the future. So let's stop talking about it and fighting over it." To underscore his conviction, Jack renewed a pledge to invest one billion yuan into the business each year for a decade. "And if after ten years we still cannot make it, we can have this discussion then."

The chatter quickly died down. The founder had spoken.

This was a serious challenge, both technically and strategically for Alibaba. It would take years for Alibaba Cloud to prove that Jack's faith was justified. During the grueling R&D stage, Wang and his team were often roused from sleep by alarms over the latest test failure. Many engineers worked long hours, fixing the system's bugs. Some of them spent so many lonely nights away from home that they set their ringtones to recordings of their children's voices.

It wasn't until 2013 that their efforts paid off. During Alibaba's big promotional event, the Double 11 Singles' Day shopping festival, five thousand servers were connected to successfully manage another year of record-breaking transactions—all on the company's internal cloud and all without error. Jack's vision had been severely tested and questioned, but it had finally paid off and added a significant new dimension to the Alibaba ecosystem. The four-year marathon was longer than anticipated, but it significantly strengthened the company's infrastructure.

Singles' Day: The Olympics of E-Commerce

There is no greater demonstration of the Alibaba ecosystem—in all its breadth and depth—than the annual Singles' Day festival.

The first Singles' Day in China is said to have been invented in the student dorms at Nanjing University in 1993, only a few years before I arrived there for graduate school. Fed up with the pressures from Chinese society to settle into long-term relationships and marriage, students gathered on campus to declare and celebrate being single. As the "holiday" spread in the years afterward, it became an occasion for singles to self-pamper, buying themselves the type of gifts they would otherwise receive from partners. Before long, the celebration was embraced by all—regardless of relationship status—as a day of guilt-free shopping.

Alibaba seized this opportunity and announced its inaugural Singles' Day online shopping festival on November 11, 2009. The results from that first year were hardly impressive: only twenty-seven merchants on Taobao and Tmall participated, offering a range of modest discounts between the two platforms. Still, the one-day promotion managed to generate nearly $8 million in sales. No one then could have foreseen the global phenomenon it would soon become.

The boundless potential of the web touched off the appetite of Chinese consumers, and the success of Singles' Day rocketed upward. Within three years, sales from the event had grown to eclipse Black Friday and Cyber Monday combined. By 2014, with Alibaba all in on mobile shopping, more than 40 percent of the holiday's $9 billion in sales came from customers using the Taobao and Tmall mobile apps.

With his eye for dramatic flourishes, Jack counted down to the 2015 Singles' Day kickoff with a four-hour program of promotional spectacles, including a cameo by James Bond star Daniel Craig, televised to millions of avid fans. Sales that year climbed 60 percent to $14.3 billion, almost two-thirds of which came from mobile devices.

I think of Singles' Day—or *shuang shiyi* (双十一), meaning "double eleven," as it's often called in China—as the Olympics of e-commerce. While Alibaba's big event is annual rather than every four years, the comparison still evokes the element of *performance* central to both. *Performance* means two things here. First, both events can be understood as a kind of show entertaining a massive global audience. Just as Olympic athletes and their fans come from all over the world, so too do the merchants and buyers on Alibaba's platforms. In 2020, Tmall Global, the company's cross-border marketplace, featured more than 26,000 imported brands from eighty-four countries and regions.

The second feature of the comparison looks inward. In the same way that the Olympics showcase remarkable physical capabilities—how fast someone can sprint, jump, or swim—Singles' Day is a powerful measure of what the Alibaba ecosystem can accomplish with everyone mobilized. It is a commercial feat, one that requires hours of training and coordination both internally and externally with participating merchants. In fact, this was how the company began to treat Singles' Day soon after its inception—as a giant annual stress test, challenging every link in Alibaba's human and technological value chain and stretching our limits. In fact, we began to use these singular events as testing opportunities for innovative ideas. For example, in 2021 the festival shifted its focus to green and eco-friendly products and sustainable practices through recycling.

Untying the Supply Chain Snarl

Building **Cainiao** required a deep understanding of Alibaba's systems and how they linked together, the nuts and bolts of the entire Alibaba ecosystem.

Cainiao increased the efficiency of China's logistics system.

Nevertheless, every decision has its trade-offs, and as much as we eventually benefited from the cloud system, it meant we made fewer investments and less progress in another critical growth area, mobile device applications. As a result of the single-minded focus for those years on the cloud architecture, we ended up having to play catch-up with one of our large competitors, Tencent, on the use of mobile apps among Chinese netizens.

Tencent, Alibaba's contemporary from the nascent days of the Chinese tech sector, launched its mobile messaging service, WeChat, in 2011, just as the central government's twelfth Five-Year Plan called for championing China's rise into global e-commerce leadership. WeChat's growth was extraordinary, reaching three hundred million users in less than two years. In 2013, when Tencent unveiled WeChat's own in-app payments service, WeChat Pay, there was no doubt that Alibaba would be affected.

It's an advantage that has not gone away. To this day Alibaba continues to struggle for the time and attention of those mobile users because Tencent has built up such a large captive audience through the popularity of WeChat. It has proven very difficult to win over those users and thus get them to use other Alibaba apps in the company's ecosystem.

Mobile shopping *was* taking over. In 2009, the number of Chinese mobile users hit 233 million, roughly double the previous year. Within four years, the number surpassed the half-billion mark. Because of falling costs and wider access, many people across the country were finally able to navigate the internet easily through their smartphones.

Alibaba was still the dominant force in China's e-commerce sector, controlling 80 percent of the market at the time, yet concerns were mounting about the company's failure to compete in the shift to mobile interface. "If Alibaba did nothing, WeChat could completely take over mobile commerce," warned a Reuters article.

Jack rallied the troops again for a major initiative. In 2013, when it was already clear that the company faced a major problem, he announced a company-wide all-in campaign to seize the lead in mobile and its many services. Within months, the company announced major investments in music streaming and ride hailing and a whopping $600 million stake in Sina Weibo, China's Twitter-like microblogging site.

But these diversification efforts were secondary to the focus on upgrading the company's core e-commerce businesses for the mobile age. Jack emphasized what was most crucial: rebuilding Alibaba's online marketplaces to be more user-friendly on mobile devices, including the creation of a special in-house team to accelerate the launch of Mobile Taobao, a dedicated app for the flagship C2C marketplace.

But Alibaba Group was not only responding to the fast-moving mobile challenge that year: we also found ourselves contending with a far more mundane headache—logistics and the challenge of last mile delivery. For instance, on Singles' Day in 2011, we managed to generate more than five billion yuan in sales, but the volume of shipments overwhelmed the Chinese postal service, and it took them months to finish delivering all the packages ordered on that day. Logistics was playing catchup to the speed of the internet and losing badly. During the 2012 Singles' Day, sales nearly quadrupled to nineteen billion yuan, but that, too, quickly crashed the overwhelmed postal system.

By the end of 2013, the first 4G networks were rolling out in China, and the push for mobile commerce was paying off. Alibaba's percentage of China's fourth-quarter mobile merchandise sales nearly tripled compared to the year before. But that volume was causing continual breakdowns in the logistics chains, holding back the benefits of our expanded mobile sales volume. We seemed to be caught in a vicious cycle, with electrons in a losing battle against old-school delivery vans.

The answer was Cainiao.

The Secret Fuel for Delivery Vans: Data

In Chinese, *cainiao* is a slang term for "rookie" or "newbie," an appropriate name for a venture taking on the logistics snarl in a way that perhaps only a rookie would. Leading the effort was Tong Wenhong (also known as Judy Tong), who, when it came to the logistics industry, was a newbie herself. Tong had begun her Alibaba career in 2000 as a secretary, handling phone calls to the tiny front desk, but quickly proved quite capable.

She wisely used her experience staffing the phones to teach herself about handling and responding to customer needs, and her expertise soon caught the attention of Lucy Peng, one of Alibaba's cofounders and the company's first chief people officer (CPO). As Alibaba grew, so did Tong's career trajectory. Lucy moved her to the company's growing administrative division, then made her the department's vice president.

Building Cainiao required a deep understanding of Alibaba's systems and how they linked together, the nuts and bolts of the entire Alibaba ecosystem. The focus needed to be on details and knitting them into a smoothly functioning operation, combining existing elements with new capabilities.

Tong was untested in the logistics chain itself, but she had a deep understanding of the company's operations. And Jack felt that, as with Jonathan Lu, the former hotel manager chosen to lead Alipay, Tong's unconventional background was an asset, since she would not be bound by old ideas and might provide unique solutions.

Jack's goals for Cainiao were, to put it mildly, audacious: within ten years, he wanted to establish a distribution network that could guarantee delivery of a package anywhere in China within twenty-four hours and anywhere in the world within seventy-two hours. But his basic strategy was not to just throw more resources at the system and increase its size: he wanted to rethink and streamline it.

He even announced publicly that he intended to reduce logistics costs nationwide from nearly a fifth of the country's GDP in 2012 to less than 5 percent. The mechanism to achieve all this, he resolved, would not be more facilities, more vehicles, or increased numbers of personnel. It would be data.

An analysis determined that gaps in information about consumer behavior and shipping patterns were creating uncertainty in the logistics system and adding to costs. Logistics managers were forced into being reactive and struggled to plan how much or how often they should transport merchandise between locations or deploy their vehicles and other resources. One store might quickly sell out of an item, while another hardly sold any at all. As a result, the analysis found, trucks motored between warehouses unnecessarily, frequently returning with empty containers. These were just a few of the inefficiencies that big data could address.

Unlike in the United States, where even the biggest online retailers were long viewed as supplementary to ubiquitous brick-and-mortar stores, most Chinese have never had access to a mature retail sector. There was no extensive system in China of shopping malls or big box retailers. When Alibaba and other e-commerce websites entered the picture, it unlocked entirely new levels of consumer zeal, one of the reasons that the delivery networks were so strained.

The size and enormous spread of China's population also presented challenges. More than a hundred cities have populations of a million or more, compared to only nine in the United States. With Alibaba generating close to one hundred million packages a day, we calculated that at least two million delivery workers would be needed to process everything. Never had a logistics company been created from scratch to satisfy such demand. There had to be another way.

Cainiao began by focusing not just on the speed of delivery but also on the *distance* that packages needed to travel. For a country as large as China, Cainiao figured out that what mattered more than simple traffic flow was how close the merchant was to the customer making the purchase. In this age of big data, extraordinary results are often a matter of identifying the right variable to optimize, and Cainiao determined that this was the critical variable. Cainiao developed algorithms to check customers' queries first against the merchants within their vicinity, matching them when possible to the stores around the corner.

It became clear that, in applying these analytical tools, Cainiao was far more than a souped-up delivery service or logistics company. Reflecting the essence of the digital economy at work, Alibaba devised Cainiao to serve as a powerful, centralized "brain," capable of crunching vast amounts of data to orchestrate and amplify what other logistics firms could achieve. As the hub in a growing nationwide logistics network, Cainiao mobilized third-party providers and equipped them to efficiently fulfill orders, with the benefits ultimately going to Alibaba customers.

Importantly, Cainiao's ability to function as an exceptionally fast data platform would not have been possible without the enormous power of Alibaba Cloud. Those synergies in Alibaba's ecosystem were critical. Though Cainiao's impact has fundamentally transformed how logistics are run in China, most of its employees are engineers and data scientists, not couriers. This has been the key to Cainiao's success. Meanwhile, Tong Wenhong's career flourished alongside it. She went on to become Alibaba Group's CPO.

Silicon Valley Discovers the China Model

In March 2019, Facebook CEO Mark Zuckerberg posted a 3,200-word statement on his public profile announcing a major shift in the social media network. After more than a decade of growing Facebook by developing a complex of social apps and advertising, he announced that the company would narrow its focus to what he described as a "privacy-focused communications platform." It would center on private messaging and payments. Although there was not a single mention of China in his post, tech industry analysts were quick to point out that his vision already existed—in the form of Tencent's super-app WeChat.

Whether he publicly acknowledged it or not, Zuckerberg was not the first to embrace innovations from Asia—particularly in e-commerce. The first Amazon Prime Day in 2015 was a clear descendant of Alibaba's

Singles' Day, while Amazon Live, the US retailer's livestreamed shopping channel, was launched after the tremendous success of livestreaming on Taobao and Tmall. The list goes on: Snapchat introduced Snap Games, its in-app games, after seeing how popular Mini Games had become among WeChat users; Instagram modeled its in-app shopping feature after the Chinese social media and shopping platform RED, better known as Xiaohongshu, or Little Red Book, in China.

According to Gartner, the financial research firm, although Chinese tech firms had previously developed as copycat models—think "China's Google" for Baidu, or "China's eBay" for Taobao—the flow of innovations has in some instances reversed.

"These days, it's often the US companies that are the ones late to the game with successful new features," Gartner wrote.

The willingness of China's tech companies and consumers to try new things, particularly China's nine hundred million smartphone users, has produced a dynamic testing lab, allowing the country's entrepreneurs to accumulate data and create new products and applications at an unmatched pace.

The Data Revolution and the Chinese Ecosystem

In the current era of the fourth industrial revolution, data is the fuel for change, the lifeblood of invention. Yet unlike the fuel of the previous stage of the industrial revolution, notably oil and gas, data is renewable and nonrivalrous. Instead of *depleting* data by using it, *more* is produced through that usage. It can be replicated and shared at nearly zero cost. That releases companies from the zero-sum mentality of the past, a profound shift. Producing and sharing data among groups, organizations, and companies creates greater value for everyone.

Alibaba's ecosystem provides a preview of what is possible in that digital world with new dimensions and opportunities. With easy access to aggregated market and industry data, merchants of all sizes can better analyze and prepare for future demand, streamline manufacturing processes, create new products, and examine customer demographics to target their advertising.

Third-party logistics providers can identify patterns in delivery demand, allocate their fleets to the most efficient locations, and even invest in new warehouses with added confidence. How we use data will stimulate further insights, allowing companies to work smarter and more collaboratively to better serve their customers.

Jack always understood that in markets dominated by large corporations and state-owned enterprises, technology could be the great equalizer for SMEs. He knew that his real mission, the soul of his digital project, was spreading opportunity beyond the traditional elite circles. Alibaba, in his vision, would place a priority on developing a platform for these struggling entrepreneurs to thrive, compete, and spread prosperity on a far more equitable basis than China had experienced before.

Alibaba Ecosystem

The Alibaba ecosystem.

Solving Real Problems

Shares of Alibaba did what stocks so often do in their first months of trading. Initially they soared, then declined as China's economic growth slowed. Some commentators piled on and declared the gold rush over. I still remember the cover story of *Barron's*, the investment magazine, published a year after the IPO, "Alibaba: Why It Could Fall 50% Further." Whatever the short-term ups and downs, the reality was that China itself, not just Alibaba, had arrived at a new level of economic influence. Jack's vision and success contributed to and reflected that rise.

Today, the Taobao universe supports more than two billion products, and more than three hundred million people use Taobao's mobile apps—on average seven times each day. Alipay supports more than one billion users across the world, including fifty million SMEs. Alibaba Cloud coordinates more than one hundred thousand servers, providing digital services that support millions of organizations around the world. In 2020, it served nearly four out of every ten companies on the annual Fortune 500 list. And thanks to Cainiao, some people are receiving their Taobao orders as quickly as ten minutes after they click "buy."

Ultimately, few companies can currently match Alibaba's end-to-end coverage of entire economies, its ecosystem. Will Alibaba be able to maintain such an ecosystem in light of the shift in the local and international regulatory environment? That remains to be seen. However, so long as the company can remain focused on its original purpose of solving societal problems while contributing to inclusive growth, it shall remain resilient and adaptable.

In a talk for California Institute of Technology in 2021, Daniel Zhang, by that time Alibaba's chairman and CEO, explained the company's evolution. Initially, he said, Alibaba was a one-dimensional e-commerce company. But that changed in 2009, when Alibaba began investing in other technologies and applications, particularly cloud computing.

"Now we position ourselves as an *infrastructure service provider*, including commerce, logistics, financial services, and cloud computing," he said. The original mission—making business easy to do from anywhere— has never changed. But today, the key to fulfilling that mission is in mobilizing magical clouds of electrons. "The digital infrastructure we build up," Zhang said, "is actually the answer."

At each stage in its life cycle, Alibaba never focused on expansion as a singularly important objective but instead on solving the real social and business problems we encountered.

Part II

The Secret Sauce

Part II

The
Secret
Sauce

4

East Meets West

Finding the Tao

One afternoon in early 2016, Jack called me into his office. He never liked having a proper table in there, and I found him reclining leisurely across his big white couch.

"Brian, let's talk about some ideas I have for you..."

He sat upright and looked me in the eyes. "What do you think about helping me create a new school? I'm talking about a unique internal training program to prepare our company for globalization."

"You see how much more work we need to do around the world," he continued. "I'm worried that we don't have enough people to fill these jobs. You're one of our company's longest serving foreign staff. You're the international guy who understands the company culture best. I think you should lead this program as your next role!"

I had been contemplating a change for a while. After two years as Jack's chief of staff, administering everything from speech prep to diplomatic protocols with heads of state and charitable initiatives, my impulse was to return to the business side. I looked forward to a more independent role, where I could make operational decisions and build something.

Jack continued describing his idea to me over the next couple weeks, and it began to sink in just how long he had been contemplating it. For all of Alibaba's triumphs and milestones as a soaring business, Jack, in his heart, never stopped being a teacher. A school was the perfect idea.

Jack's plan said a great deal about one of his most heartfelt principles—that the key to building a highly successful enterprise was not just developing leaders with the standard knowledge of management and financial skills but also focusing on their spirit, their sense of mission, their commitment to meeting society's needs more than just KPIs.

The academy Jack hoped to create would be a highly selective internal training program for instilling Alibaba's values into its next generation of the company's international leaders. That, he believed, was the secret to continued performance.

We were in the middle of a major overseas expansion at the time—opening offices in Southeast Asia, India, the United States, and Europe. We needed to make sure that our overseas operations possessed the same spirit, vitality, and distinctive way of seeing the world as our China-based operations. These operations were not just pieces on a chessboard but also extensions of Jack's vision and the desire to share his sense of purpose, his social perspective, with the international markets.

By July 2016, we had finished drawing up the plans for our first class of what came to be known as the Alibaba Global Leadership Academy (AGLA). The first class of the twelve-month rotational program received more than three thousand applicants, from which we had to make the difficult choices for selecting thirty-two participants. We would work hard to teach this group the Tao of Alibaba.

Learning the Way

As a term, *tao* is a popular abstraction that has trickled into popular culture with varying levels of seriousness—everything from self-help books on mindfulness to Hollywood screenplays to trendy Asia-themed bistro clubs in Las Vegas and New York.

In truth, it is challenging to really understand. By its very elusive nature, this philosophy defies a clear, straightforward definition. In

Chinese, tao (道) translates literally to "way," in the sense of how one proceeds or conducts one's affairs. It is centuries old, more a philosophy than a religion, that attempts to describe the relationship among things within the universe and how people can find a useful and satisfying path in their lives. It focuses on guidance rather than faith. For the purposes of this book, I refer to tao as a depiction of the way, the path, or the essence of something.

What we described as the Tao of Alibaba was the secret sauce that powered the company's success. Like the tao of Chinese philosophy, it resists a concise encapsulation in words, and it is not a mere listing of concepts. Although it cannot be boiled down to a handful of pithy expressions, easily copied from a chalkboard, that does not mean it cannot be grasped and employed. Experience is the best teacher. And a key tenet of the philosophy is learning through action and self-reflection. Hence, Part II of this book reflects my appreciation of the value of this tao based on my experiences at Alibaba.

Embracing Contradictions

Jack Ma stands out for the way he has always trusted his own firsthand experience, his own observations and gut instincts, what he sees with his own eyes and his native understanding of people and their yearnings, more than the abstractions of textbooks or principles taught in classrooms. He does not want to do things a certain way because everyone else follows that approach. Instincts are better than predetermined plans, in his view. He prizes spontaneity.

Perhaps the best example of this was when, years ago, the moderator at a Harvard Business School conference asked him about the secret to his company's success. Jack being Jack, he grinned and replied: "Alibaba succeeded because we (1) had no plan, (2) no technology, and (3) no money."

Jack by then had already established himself as a charismatic personality with a flair for outrageous statements. Yet his seemingly sarcastic reply left many in the crowd scratching their heads. Beneath its confounding veneer, however, the statement accurately reflected a core mind-set that Jack instilled in the Alibaba culture, and those who were really listening closely heard it.

(1) Planning shouldn't trump doing.

Jack's handling of the "business plan" question always befuddled observers. "We have no business plan" was his preferred response. Occasionally, he would also tell audiences that "change is the best business plan."

The truth is that while Jack did not fancy PowerPoint presentations or complicated charts to make his business case, he made his objectives clear and then enabled us to come up with creative approaches to solving the problems. In Alibaba's early days, he encouraged us to spend more time executing, rather than having endless meetings where we simply talked about executing.

Jack's approach reflected the rapid pace of the internet era. Instead of spending all day crafting the perfect business plan, he preferred we implement ideas, iterate, and then implement again whenever we encountered failure. While understandably some would argue this wasn't the most efficient approach, it enabled us to "move fast and break things," an approach that worked for a small scrappy start-up.

(2) Because Jack was not a software engineer, we focused on how the platform worked for our customers, not coding accomplishments.

As a former English teacher, Jack did not possess any specialized computer skills. He could not code or talk knowledgeably about software architecture. This perceived deficiency appeared even more limiting when contrasted with the founders of competing Chinese internet firms, many of whom had trained at top American programs and had advanced degrees in computer science and engineering to show for it.

What Jack did understand, however, was the mentality of Alibaba's target customers. Many of these factory bosses were the same age as him. They had similar backgrounds, had endured the same life struggles in a China that had undergone enormous change. They had a similar understanding—or lack thereof—of IT principles.

Jack saw himself, in fact, as sort of an everyman. He often said, "If Jack doesn't understand how to use his team's products, then most likely none of the other small and medium-sized business owners will either!" He came to describe himself as Alibaba's CTO—chief testing officer— whose primary role was to ensure that the firm's technology was so user-friendly that all those seat-of-the-pants entrepreneurs could employ it.

Jack's relative lack of technical expertise also meant that he had no choice but to trust in his engineering team. Ambitious business leaders are often reluctant to give up control of any important functions, but Jack knew from the start that he needed to rely on the expertise of others. Instead of micromanaging his engineering staff, he gave them the space they needed to develop and innovate. He would set a broad direction and then signal his confidence in the team that would get us there. His trust in others paid off: he could steer the company effectively without getting in the way himself.

(3) Having no money in the early days actually made the company more resilient.

One of the biggest mistakes that well-funded companies make in their initial development is wasting cash because they have a surplus of it. Jack used to frequently tell us: "Money makes you stupid." This wasn't just a soundbite—it was a hard-earned lesson. Early in Alibaba's history, a mistake nearly toppled our company into bankruptcy. But it drove home the lesson that money should not be the solution for every problem.

In 2000, we decided to hire a sterling 4A agency—belonging to the American Association of Advertising Agencies—to furnish all the marketing bells and whistles that any large corporation would have. Soon we had outsourced most of our creative, production, and media-buying functions to them. After a year of burning through a sizable portion of our company funds on this surge of new marketing, we woke up to the devastating fact that we could perform many of the necessary tasks in-house—and often with better results. We switched to designing our own logos, writing our own taglines for advertising copy, and launching guerrilla marketing campaigns that utilized nontraditional media and unconventional tactics to bring attention to our brand. We were able to recover, but it was a costly mistake.

Having limited financial resources pushed us to become more innovative in how we approached building our business. By the time we had to compete with international powerhouses like eBay, we had already honed our skills for guerrilla marketing warfare in a way that repeatedly caught our competitors by surprise.

Tai Chi Management Principles in Practice

Jack talked often about leading a company that infused Western management with Chinese philosophy, a blending of East and West. While many of Alibaba's founding principles are Chinese, it utilizes Western business and management systems, combining them in its own unique formulations. Once inside the company, constant dichotomies must be addressed, including nurturing the small to grow and become large in aggregate, balancing quality versus speed, becoming strong yet supple, figuring out how much time to spend in planning versus just being responsive.

Such concepts—and what they look like in practice—might appear on the surface to involve a series of contradictions. For example, Alibaba is a mission-driven company, and even in meetings with the company's leadership, we rarely discussed profitability or other financial metrics. We talked about whether we were serving our customer base well and if the benefits of the Alibaba platform were being shared widely. In fact, during the 2008 financial crisis Alibaba launched three campaigns, Dark Cloud, Wild Wind, and Spring Thunder, which consisted of slashing subscription fees for existing services and offering new free services to help millions of SMEs join the Alibaba ecosystem and survive that economic peril. More recently, during the COVID outbreak, Alibaba launched initiatives to help mom-and-pop stores and restaurants, enabling delivery services amidst lockdowns, waiving some service fees, and offering interest-free microloans to SMEs in need. We never lost sight, however, of our need to be profitable to survive and continue to grow.

Similarly, one of Alibaba's core strengths is an unshakable set of values that, together, serve as our strategic North Star. The values, however, are not static. During my final months at the company in the spring of 2020, we implemented six new values—the third such makeover since I first joined the company in 1999. Even as we continue to prioritize the intangible values that, we believe, set Alibaba apart, that make us an indispensable partner to and champion for our customers, we remain open and ready for change—because the world never stops changing. Today, our company values reflect the changes we were experiencing in a vastly different society and industry than the one we encountered when Jack started Alibaba.

Alibaba's leaders engage with and navigate this changing climate and many apparent contradictions through what I call the tai chi management model. Westerners may be surprised to learn that the teachings of a venerable Eastern tradition permeate a high-tech company. But it has been an extremely useful guiding philosophy in a world filled with contradictions. Embedded in the philosophy of tai chi is the idea that the world around us consists of both yin and yang: it is simultaneously heaven and earth, full and empty, an ongoing series of opposing and interlocking pairs, an ongoing dialectic.

In 2009, for example, as Alibaba approached its ten-year celebration, Jack gathered the company's senior leaders for a talk to reflect on how we got there. One of the most heated discussions focused on how our management should treat the notion of *du* (度)—roughly translated as the extent or degree to which a course of action should be pursued.

"In its early days, a start-up has only two choices. Black or white? You can choose only one," Jack said. "When there are only a few dozen people in your company, you walk in either this direction or in that direction. It's that simple."

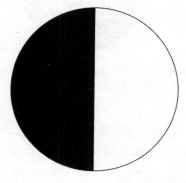

New companies have simple black-and-white choices.

Yes, Instead of Black or White

In 2002, Alibaba had hired Savio Kwan, a seasoned multinational executive, to be its chief operating officer. As the former head of General Electric's medical equipment division, Savio had introduced the company to his "yes" theory. As Jack explained this, "If you need to decide, should I hire top performers or those who adhere to the company values? Savio says, *yes*. He says we will hire people who can follow directions *and* who can get things done."

Employees at most companies will resist such an idea, operating under the assumption that one must decide on a single path, not an all-of-the-above approach. Not Jack. "This is the reason why Alibaba is more successful today than other companies," Jack said. "We prioritize values *and* we prioritize achievement. For the last several years, we have insisted on combining these things together. Everyone always says you have to choose. East or west? Black or white? We say: yes and yes."

Embracing the yes approach remakes the organization's composition so that it is more appropriately represented by the yin-yang (阴阳) symbol.

The yin-yang symbol: more dynamic decisions that strike a seamless balance.

The image above represents the dynamic relationship between dualities, in perpetual movement. The ideal state is referred to as the mean, the point of balance where there is the precise amount of each element to enable a seamless flow.

Finding that mean is the eternal test facing everyone. Every company needs to find its own balance between start-up energy versus corporate stability, established MBA standards versus entrepreneurial hacking, a Chinese-style market entry versus more conventional Western tactics.

The final tier of the progression is represented by taking the yin yang diagram and adding another element—placing a black dot in the white field and a white dot in the black, opposing yet unified, different yet interconnected. This forms the *taijitu* (太极图), or the tai chi symbol.

The tai chi symbol or taijitu: *hardest of all is infusing an element of each into the other.*

The practice of tai chi is the pursuit of harmony and balance, the embrace of intuition, and the endeavor to understand what might be hidden or concealed and the recognition that these pursuits are just as important as logic and reasoning. Being in such a state is what enables flashes of inspiration, new ideas that can lead to true innovation. Tai chi teaches that truly comprehending any situation and achieving those breakthroughs requires appreciating more than just what is visible.

I came to see that in understanding and embracing these special, usually highly effective qualities I had to unlearn some of the management practices I had absorbed during my years in the United States. These were not easy beliefs to shed, and the process of leaving them behind was, at times, humbling.

A great example of this process occurred when Yahoo! in 2005 bought a 40 percent stake in Alibaba for $1 billion. I initially regarded it as a moment of triumph, and we celebrated what seemed like a milestone: we had grown up and evolved into a more stable, mature operation. We excitedly discussed all the collaborations, partnerships, and new investments to come and began a wave of new hires—people with impeccable qualifications—to responsibly guide the initiatives. No more daily improvisation, I thought: Alibaba was finally going to modernize. But a couple of years later, I learned a far different lesson from that milestone. Nearly all those impressive new hires of industry-tested veterans—bearing the "right" credentials from the "right" institutions—had either left or been fired. I had gone through two different bosses—both respected in the industry.

We discovered that some of the conventional practices in running a business did not work well in the Chinese digital marketplace or in the challenging e-commerce business. I had been to business school, where I had been drilled in the application of KPIs and analytical decision-making, but I realized that in many instances I needed to abandon those boxy, narrow formulas and do a better job of following my instincts. Not everything can be measured by an equation.

Today, I also see clearly what a huge contrast there is between the rigid, conservative culture of many established corporations and the more energized metabolism of a start-up like Alibaba. Despite Alibaba's tremendous growth, we clung to our self-image as the scrappy, nimble challenger. Change was a constant in all that we did—from strategy and planning to the company's organizational structure. It felt at times that we were announcing major reorganizations every six months.

We built new products and workflow processes only to tear them down to keep up with the market and to prevent old habits from obstructing new methods. Sometimes we created two departments and assigned them the same task, just to see who could do it better. It was trial and error, and only the best performers survived.

At the same time, executives at Alibaba, including myself, devoted a considerable amount of time to assessing nonoperational issues, such as our company's values and culture, when it would have been far easier just to push forward and focus on the business tasks at hand. But, we found, this extra effort paid off in the long run since it strengthened our culture, and that produced better decision-making. This approach contradicted many business school models, but little by little, I came to appreciate how it formed the system of tao that permeated Alibaba, contributed to our agility, and enhanced our performance.

The Sage Ruler

The tenets of Taoism are described in a text called the *Tao Te Ching*. It is commonly regarded as the work of the great Chinese philosopher Laozi (also known as Lao Tzu) who reputedly lived before Confucius, but scholars debate whether he was a real historical character or represents a collection of ideas from numerous philosophers in the Taoist tradition.

Its principles hark back to an era known as China's Warring States Period (475–221 BCE), a spasm of history marked by violent conquest and turmoil. In this environment, the *Tao Te Ching* was primarily concerned with achieving and maintaining a stable, tranquil society—the chief preoccupation of many philosophical texts at the time. Some *Tao Te Ching* principles were written explicitly to a "ruler," while many others addressed the "sage."

This was not an inconsistency: in the ancient Chinese worldview, the ideal ruler *was* a sage. Today, we take a related perspective. Effective business leaders, rather than managing with force, status, and raw power, should rule with a far more valuable and effective quality—wisdom.

In contrast to some Western traditions, Taoist philosophy is concerned less with breaking through to some absolute truth and more with cultivating effective strategies and habits for achieving a state of harmony. One fundamental tenet of this philosophy is to treat things as interconnected rather than discrete, differentiated entities. Seeing these connections guides people toward a more harmonious life path, more in line with the natural way, or tao, of life.

In the modern business world, strength and domination are typically revered, making it difficult to appreciate how some of the tenets of Taoism apply. In their book, *The Path: What Chinese Philosophers Can Teach Us About the Good Life,* Michael Puett and Christine Gross-Loh stress this sense of interconnectedness as a key to real power. The *Tao*

Te Ching, they wrote, "derives from appreciating the power of seeming weakness, understanding the pitfalls of differentiation, and seeing the world as interrelated. Rather than think that power comes from strength prevailing over strength, we can understand that true power comes from understanding the connections between disparate things, situations, and people. All of this comes from an understanding of what the *Laozi* calls the *Tao*, or 'the Way.' "

In this view, finding ways to reinforce the interconnectedness within the world brings us closer to the way, a type of harmony with natural forces that allows us to harness that power to shape the outcome of events and situations. Understanding how to do this gives us the ability to realize new possibilities. As Puett and Gross-Loh say, "we can actually gain the power to re-create the Way all the time, at every moment."

Work Is Life and Life Is Work

For Jack, this was reflected in the fact that he tried not to separate his thinking between the professional and the personal, work and leisure. He would often say, "At Alibaba, work is life and life is work." He did not mean that everyone needed to be thinking about work all the time, but that what you experience and learn through your job should be applicable to everyday concerns and some life lessons should be applicable to work. They are connected. When framed in this way, there is no longer a need to "balance" work and life, as both activities should be mutually reinforcing, each improving the other.

This belief is reflected in the latest iteration of Alibaba's sixth value: "To live seriously, work happily." (Alibaba's six values as a whole will be further explored in Chapter 5.) The sixth value encourages employees to think seriously about what they want to achieve in their lives and how work fits into that plan. Work can thus become more fulfilling as it helps satisfy these broad life objectives.

Puett and Gross-Loh write about Taoism's stress on avoiding "false distinctions." They state that "by dividing up life and believing that these aspects of our lives are unrelated to one another, we restrict what we are capable of doing and becoming. The *Laozi* would say that not only are mystical enlightenment and our everyday lives related, but that by separating them, we have fundamentally misunderstood both."

Nonaction as a Call to Action

Another concept central to Taoism is *wuwei* (无为), meaning "nonaction" or "nondoing." But this is not a call for passivity. *Wuwei* calls for abandoning compulsive and contrived impulses and letting one's behavior align with the more natural flow of life, the organic unfolding of nature. The simplest example is a tree: it grows as part of a natural and inevitable process, not an assertion in opposition to some force.

Put in more common terms, this seems to be about getting out of your own way in making decisions, to try and look broadly at situations and understand how you can fit into them and achieve your broader goals. In my own case, I confronted this challenge before I accepted Jack's suggestion that I establish the Alibaba Global Leadership Academy. I had been in talks with the management team at a division called Alibaba Sports, a new business line that was looking for someone to lead their international efforts. The role checked many of my boxes at the time, and I was eager to pursue it.

In hindsight, it would have been shortsighted if I had insisted on that opportunity. I would have let down my mentor, Jack, who had selected me for a job that he regarded as a major priority. I would have been serving a narrow corporate interest rather than a job that related to our broader mission. And most importantly, in hindsight, not accepting this role would have led me on a detour from the ultimate path of seeking a pursuit that would allow me to positively impact the greatest number of people.

Wuwei suggests being aware of and acting in accordance with those broader currents rather than just exercising will and seeking some near-term aim. This reflects Taoist principles. *Wuwei* calls for identifying the direction of greatest energy and connecting yourself to the flow.

The Tao of Alibaba was an expression first coined by the company's former chief strategy officer (CSO), Ming Zeng. It captured our belief that Alibaba's strength was not in developing and pushing a product on the market but in truly grasping the needs of its customers and harmonizing its teams and operations to better align with and serve those needs, common human desires, and goals.

Components of the Tao of Alibaba.

Creating Replicable Success

Jack's concern for these sometimes abstract philosophical issues was not idle daydreaming. It related to his overriding interest in constructing a solid base of thinking, a culture, that could guide a business constantly growing, evolving, and changing. Jack liked to point out that there are people who often speak loudly of their ideas for the industry and the future. But if they don't have the ability, or techniques, to turn their ideas into reality, their talk amounts to nothing more than empty theory. Meanwhile, there are others who have collected a lot of impressive results yet have no clear idea of the conceptual basis for how they achieved them.

"If a result does not even know how it was caused, why it became this result, then it can never be replicated and can never be called a success," Jack said. "On the other hand, knowing how something was made, but not knowing its significance to society, to mankind, to oneself—then it still has no meaning."

Jack believed that to overcome those dead ends it was critical to combine three principles: *tao* (道), one's beliefs or faith; *li* (理), rules or principles; and *shu* (术), techniques or tactics. In the Tao of Alibaba, mission, vision, and values belong under the tao principle, strategy aligns with the *li* principle, and organization and people as well as performance management fit with the *shu* principle.

The starting point is mission, a synthesis of *tao-li-shu*. Jack often said that a mission is like a dream, something that gives people hope that they can achieve their loftiest aspirations. But since a mission is usually abstract, it must be articulated through a *vision* that establishes a concrete set of goals. An example of this is an organization's five-year and ten-year plans. In its first ten years, Alibaba had a vision of pursuing e-commerce. In the next ten years, its vision shifted into creating a business civilization—a vibrant ecosystem built on principles of openness, collaboration, and data sharing.

Today, more than twenty years after its founding, Alibaba's vision is to help realize a new digital economy that is inclusive to all, especially in emerging markets, providing the digital infrastructure to enable this transformation. Once the mission and vision are in place, an organization must then formulate a code of conduct—its *values*—that can motivate performance and define accepted behavioral norms.

Next comes establishing a strategy. This incorporates the mission and vision into a statement that establishes the *who, what,* and *how* of a plan. With this strategy in place, the company can design an organizational structure and hire appropriate talent to realize the goals. Proper incentives and performance management practices must be implemented to keep employees motivated and working toward the objectives.

Assembled correctly, all of these feed into a virtuous cycle: guided by the correct values within a proper structure, an organization's talent can accomplish the company's vision and ultimately fulfill its mission.

The Alibaba leader plays the role of a conductor who coordinates these four components—mission, vision, and values; strategy; organization and people; and performance management—and integrates them into the smoothly functioning, skillful management of the team.

But learning the Tao of Alibaba is an experiential exercise in addition to an intellectual one. Jack often said: "Most people must first see, and then they believe. At Alibaba, we believe and then we see."

That kind of confidence is a crucial ingredient for any venture in its early stages. Most people will wait for others to do something before they recognize it as possible. A true leader must somehow conjure that faith from day one, to motivate the enterprise. The Tao of Alibaba is, in essence, how Jack and his teams turned that faith into reality.

5

Mission, Vision, and Values

Why Are We Here?

Jack stood in front of the room and clapped his hands. "OK, everyone. Time for our evening meeting."

It was January, 2000, my first month in Hangzhou. Alibaba's high-energy leader was holding court inside our "headquarters"—one of two apartment units we occupied in a residential compound at the edge of the city's well-known West Lake. In one apartment, Jack, Lucy Peng, and a few other leaders worked alongside the firm's engineers. In the second, the product, design, and customer service staff worked. That's where I was sent, assigned to a small desk that had just been cleared.

At Jack's call, the staff gathered in what was once a living room, huddling on couches and mismatched chairs in a rough circle. Most of the thirty or so people in the room had notepads on their laps, waiting to hear what Jack had to share.

"I have great news today," Jack began. "We've finally raised money from some international investors that will allow us to become the global company we want to become!"

Cheers went up around the room.

"We have officially closed our $5 million funding from Goldman Sachs! We will start hiring new staff in Hangzhou and set up our new office in Hong Kong so we can build a company that is as strong as Yahoo or eBay! It's time to let the world know that Chinese companies exist. Our mission is to help the world do trade with China! And we will do it using the internet!"

Those were big words for such a small company. It was wonderful news, but I couldn't help wondering if expectations hadn't gotten out of hand. It was already clear that my young colleagues had a strong work ethic, but they hardly brought world-class training or experience to the table, something I thought would be essential to challenge the titans of Silicon Valley. Those companies hired talent from the best universities in America and had ready access to financing and other resources.

Our ragtag bunch consisted of some of Jack's former students, a few English teachers, some trading company employees, a journalist, and IT graduates fresh out of college. Few had internet experience, and only a handful had traveled outside China. How were we ever going to beat the dominant global tech companies? We didn't even have room for all the chairs in our "office."

Jack finished his celebratory speech by repeating his most immediate goal for tiny Alibaba: "Today we have thirty thousand members. In the next year, I want us to aim to achieve one million members. We can do it!"

Plausibility had no place in that conversation, I thought. Everyone's enthusiasm flowed from their strong identification with the company's mission and Jack's vision, and if it was part of our mission, all believed, it was certainly achievable. In the coming months, I found this collective belief system infectious. It wore down my initial skepticism. In fact, in time, I would grow to better understand and more deeply embrace Alibaba's mission, vision, and values and the way they motivated our people.

When Jack concluded the announcement, everyone scattered back to their desks, and I headed to move into a company apartment I was transferring to, following my short hotel stay. It was a ground-floor unit nearby, fronting a busy road. Peeling paint drooped from the walls. Beyond the basic amenities, there were hardly any furnishings.

I sat down on my new bed to soak it all in. Just then, a construction truck roared by my window, kicking up a swirl of dust that streamed into my room through an open vent. I covered my mouth in a fit of coughing.

The Moonshot Mentality

The first wave of Alibaba employees generally shared Jack's idealism and, to put it simply, his naïveté. Jack had charisma and a powerful ability to motivate his team to imagine an intricate, expansive future that had no apparent connection with what we saw around us. We used to joke that an important trait we all shared was being *hen sha, hen tianzhen* (很傻很天真), or in English, "simple and innocent." We believed that anything was possible if we just gave it our best efforts.

Despite our optimism, reality crept into view from time to time. Some nights, I would look around the spare furnishings in my apartment and wonder how I ended up here. I questioned whether I was making the most of my life. We spent countless all-nighters here on projects that would sometimes be shelved at the last minute, as in the DIY website builder that was never launched and other schemes that required lots of work but then failed to materialize. Sometimes, it felt like our million-member target was more than a moonshot. Was it remotely possible?

Jack's personal energy carried some of us past those moments of doubt. But having a vision, while important, only goes so far in the practical day-to-day world of business. At some point, it must be implemented with a plan. That's what Savio Kwan, Alibaba's first COO, did, codifying Jack's ideas into our company mission, vision, and values. Savio can still point to the exact moment when, for him, the breakthrough occurred.

"It was 2001, January 13, a Saturday," he recalled. "There were six of us standing around outside Jack's office, chatting away." The group included the rest of the "four O's"—Jack, the CEO, Joe, the CFO, and John Wu, the CTO—as well as two cofounders, Lucy Peng, Alibaba's CPO, and Jin Jianhang, who would later serve as the company's president.

They decided they needed to sift through Jack's notes and past speeches to identify the most important ideas and then use those to produce a coherent mission and action plan. Their efforts would form the basis of Alibaba's core mission, vision, and values statement—the guiding force or beliefs (or tao [道], as we referred to in Chapter 4) for Alibaba's development then and now. Mission gives a company purpose and defines the reason everyone has joined together. Vision provides a window through which to view the future. Values define the principles that ground the business and direct the team's behavior. Collectively, those concepts occupy the tip of the Tao of Alibaba pyramid, the beacon illuminating the right path forward.

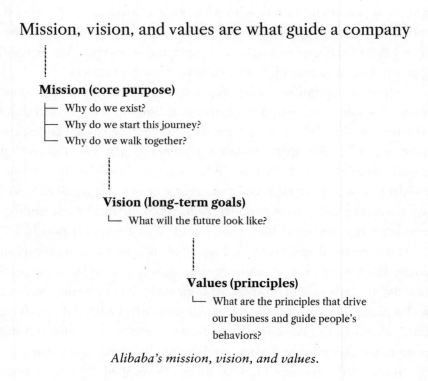

Mission, vision, and values are what guide a company

Mission (core purpose)
— Why do we exist?
— Why do we start this journey?
— Why do we walk together?

Vision (long-term goals)
— What will the future look like?

Values (principles)
— What are the principles that drive our business and guide people's behaviors?

Alibaba's mission, vision, and values.

Though we recognized the importance of each individual component, it was only through years of trial and error that we discovered how the pieces locked together to form a framework to drive sustainable and replicable development. We have since shared this knowledge with thousands of entrepreneurs and offered lessons for applying it to their own companies, in China and in many emerging markets. The Tao of

*Mission, vision, and values are the guiding principles
of the Tao of Alibaba.*

Alibaba has helped many of them transform their organizations and perform much better.

Mission: A Company's North Star

Every entrepreneur knows that the decision to launch a company is, in many ways, irrational. If the world really needed this company, then why didn't it already exist? The process of building a venture can be backbreaking, involving long hours, financial risks, lots of rejection, and limitless stress. Success is never assured. Some entrepreneurs may be driven by the wish to get rich quickly, but that is rarely sufficient to help overcome the workplace trials. For the ones who endure, other motivations usually enable them to weather the setbacks.

This is where a sense of mission enters. A company must represent something more than simply making its founders rich. Many of the entrepreneurs we admire most will say that their initial decision to create a start-up actually had very little to do with making money. And even after accruing great riches, their wealth matters little in how they manage their companies. There is usually a higher goal.

A good mission statement should explain in simple terms *why* the company exists. Alibaba's mission has always been *to make it easy to do business anywhere*. This declaration of our company's central purpose has been our North Star from day one, when we were a simple online B2B marketplace, to today, across a sprawling ecosystem of business lines and services. When somebody is hired by Alibaba today, it doesn't matter if he or she is joining Alibaba.com, Tmall, Alipay, or Alibaba Cloud; the new hire knows that his or her ultimate purpose as part of the company is to make it easy to do business anywhere, to respond to a need in the market. It is the articulation of what we, as an organization, intend to contribute to the world.

In fact, the first sentence of Chapter 8 of the *Tao Te Ching* is a line: *shangshan ruoshui* (上善若水). This phrase roughly translates as the best behavior of people in high places trickles down to all levels and nourishes all things. While it can be applied to a specific leader and his or her model behavior, it can also be applied to the importance of a company mission and how it should permeate and inspire every level of the organization and its culture to lead it toward a common purpose or goal. Hence, a mission statement is the rallying cry for all to join in a unified cause. Without a strong mission statement, other components will not follow. With a strong mission statement, the rest of the elements, such as vision, values, strategy, organization, and performance management, all flow logically.

There are several important characteristics entrepreneurs should address when writing a mission statement:

1. **It explains why your organization exists and what its core purpose is.** This is the foundational premise of the mission statement.
2. **It is clear and easy to remember.** Often, when I am conducting training and workshops for emerging market entrepreneurs, I give the participants a simple task: write down your company's mission statement. Immediately, almost everyone reaches for their laptop or smartphone to pull up their company's website to refresh their memories. Often, mission statements are too long and complicated for even the founder to be able to recite. If the founder has

not internalized the mission and can't even *articulate* his or her company's mission without a cue, how can the staff remember it?

3. **It inspires employees and customers alike.** The mission statement should effectively broadcast the company's purpose to all stakeholders. A good mission statement should inspire people to engage, learn more, connect.

4. **It is altruistic.** Too often, people hear the word *business* and think of profit-hungry ventures. All of us in the business community must challenge this narrow perception through our actions. We must ask ourselves: Beyond profits, what is it that we're really trying to achieve? An altruistic mission bends toward serving others. It looks outward, explaining what a business can do to benefit members of the larger community.

5. **It carries a long-term perspective.** The mission should not be something that can be achieved within a limited time frame; in fact, it may not even be achievable. Instead, it should provide a direction, a point on the horizon that everyone is working toward. This will become clearer when we discuss the differences between mission and vision, where vision typically involves more concrete time lines.

The Power of a Strong Mission

In crafting a concise and compelling mission statement, it is useful to examine the mission statements driving some of the great companies of our time. Here are global firms that have shaped their industries and most of our lives:

| To inspire and nurture the human spirit—one person, one cup, and one neighborhood at a time. | Think different. | We create happiness by providing the best in entertainment for people of all ages everywhere. |

Sample mission statements and slogans of other great companies.

- *To inspire and nurture the human spirit—one person, one cup, and one neighborhood at a time* (Starbucks). Starbucks presents itself as far more than a coffee provider. It quenches not thirst but human yearnings. This ethos extends beyond just the company's relationship to its customers, encompassing the way that it treats employees, engages partners, and even sources coffee beans. From executives to baristas to suppliers, everyone at Starbucks should take pride in working to improve people's spirits and wellness, the statement says. The former chairwoman and CEO of Starbucks China is a friend of mine, and she once told me about the corporate battle she fought to allow customers to bring their pets into Starbucks. Such an effort may not sound like part of the typical CEO's job, but in recognizing how big a role pets play in peoples' lives, she felt that such a policy reflected the mission of Starbucks to create an inclusive, nurturing environment for her customers.

- *Think different* (Apple). The company's slogan reads like a provocation, daring you to challenge the status quo. In Alibaba terms it could have very well served as its official mission statement. Its iconic 1997 "Think different" advertising campaign featured Albert Einstein, Bob Dylan, Martin Luther King Jr., Thomas Edison, Muhammad Ali, and Mahatma Gandhi—exemplars of once radical visions and bold accomplishments. For customers and potential business partners alike, Apple's statement stimulates a desire to join a larger cause, to be a part of the change that the company brings to the world. Apple is a brand I grew up with. Steve Jobs actually donated Macs to my high school in Palo Alto, and they opened up a world of graphic design, music composition, digital publishing, and other applications that supported my creative expression—and my life—and still do, to this day.

- *We create happiness by providing the best in entertainment for people of all ages everywhere* (Disney). For many years this was Disney's mission statement, oftentimes shortened to: "We create happiness." In either form, it effectively captured the organization's essence. Not long before the pandemic set in, I brought my two young daughters to Disneyland in Shanghai. It was my first time at the giant amusement park as an adult, and I was skeptical. In the end, I think my daughters and I were equally amazed. As a company, Disney oversees a vast empire of entertainment and entertainment-adjacent businesses. Despite its tremendous breadth, all its properties and acquisitions could be said to share the unified mission of creating happiness by providing the best in entertainment for people of all ages, everywhere, myself and my little girls included.

The example of Disney, however, also serves as a cautionary tale. Visit its website today and you will find that this new mouthful has replaced the old mission statement: the company's aim is "to entertain, inform and inspire people around the globe through the power of unparalleled storytelling, reflecting the iconic brands, creative minds and innovative technologies that make ours the world's premier entertainment company." This updated mission statement shows that even the most vaunted companies can fall prey to management-consultant speak. In this case, the mission statement sounds more like a statement of strategy than purpose.

This is not to say that there is only one way to think about the mission statement, nor a single approach to devising a company's vision and values. What I offer here is the path that has worked for Alibaba and many other industry leaders. A company with a clear, succinct, and purpose-driven mission statement will more likely be efficient and purpose driven itself.

The Important First Step

In my opinion, the toughest period in the start-up life cycle is the very beginning—just getting from zero to one.

Most emerging market entrepreneurs I have met and helped train through our company's teaching programs, Alibaba Global Initiatives, or AGI, were in early stage ventures. Of course, they often lacked resources, investment, and staff. That's the norm, so it is hard to imagine that a now giant digital company like Alibaba was ever a struggling start-up. It was, and its early crises are instructive.

Alibaba was not Jack's first attempt at starting a company. He had previously run an English language translation service called Hope Translation Agency, and then he launched his internet bulletin board, China Pages. When that business failed to catch on, he sold it to the local government. Alibaba represented his third attempt at launching a new venture.

Neither Jack nor Alibaba were well known initially, and yet, within the first year, Jack was able to secure the confidence of investors, including the prestigious global giants Goldman Sachs and Softbank. Something magical was starting to happen. He managed to recruit some exceptional talent—from Joe Tsai, at the time a respected private equity executive in Hong Kong, to John Wu, a pioneer in search technology for Yahoo!, and Savio Kwan, who left GE, a blue-chip behemoth, for this unknown internet start-up.

Why would any of those business stars have agreed to take such a tremendous risk? Part of the answer was Jack's sense of mission. If you ask Joe, John, or Savio, they will tell you that much of what drew them to Alibaba was Jack's enthusiasm for what real social good they could accomplish together—the impact they could have by striving *to make it easy to do business anywhere* for those tens of thousands of Chinese SMEs struggling for their own success. It's the same for Jack's other seventeen cofounders and the same for many in those first waves of employees—myself included. What had attracted me to Alibaba was how Jack articulated his purpose for creating the company.

During those early days in his Hangzhou apartment, Jack would regularly host all-hands meetings in the living room. He would share his ideas for the future, the increasing role he saw the internet playing in the coming decades, and how small businesses could thrive on these changes, creating exceptional new opportunities and prosperity, especially for marginalized communities that had long been neglected. It was our work, he said, that would help bring this new vision to reality. And he made it clear that his approach had a real chance of succeeding. When put in such terms, this did not sound like a job. It sounded like a mission.

Jack was never shy about comparing Alibaba to the world's leading tech firms in Silicon Valley because he truly believed that he could match, if not exceed, what they were achieving. We all had skeptical moments, but it was the stream of little wins that made this seem doable and that kept us motivated. If we had thirty thousand users, then why not one million?

As a senior director of our global business development efforts during the late 2000s, I spent much of my time traveling to places like India, Turkey, Malaysia, Vietnam, and other emerging markets. We were always learning, seeking to understand the real needs in the market and how we could fulfill those needs with our digital platform.

Typical was a trip to Mumbai in 2010, which my colleagues and I visited to hear customer stories. This was not high-tech glory. Each day we scheduled four or more site visits, hopping on *tuk-tuks* (auto rickshaws), winding through narrow streets packed with hawkers, laborers, and children bound for school. We met with textile makers, auto-part manufacturers, chemical distributors, jewelers, and more. Some of the offices were found up three flights of creaky stairs in dilapidated buildings, staff members crammed in, working behind overburdened desks or machinery. But we listened and brought home ideas for how we could serve them. While the travel schedule was tough and my mandate felt overwhelming at times, the thing that kept me motivated was the thrill I felt each time an Alibaba user told me how our platform had changed his or her life.

The lessons from those experiences stick with me to this day: never underestimate the power of a strong mission and the importance of being able to communicate it to your customers, your employees, your investors, your partners. At Alibaba, they all understood what we were about.

Your Vision Makes Your Mission Tangible

The mission statement codifies a company's purpose and establishes a far-reaching goal capable of inspiring and motivating others. How do firms go about achieving the goals, and how do they know if they have made meaningful progress?

The vision statement is the next line of action. It translates more abstract aims into concrete, measurable objectives. At the highest level, an organization's vision statement articulates the results it hopes to achieve and places them within a specific time frame. The vision statement describes what the company will look like at various points in the future.

Alibaba's vision statements have evolved over the years as the company scaled new heights and new targets replaced the old ones. Earlier iterations of Alibaba's own vision statement provide insight into how a company's vision can shift as it expands and broadens its perspective.

The Five Characteristics of a Strong Vision

Here are five elements that every founder should keep in mind when composing the company's vision statement:

1. **It should be aligned with your mission statement.** Does achieving the vision fulfill the company's long-term mission?
2. **It should forecast a meaningful time frame.** We often advise company founders to fix a five- to ten-year time line to their vision statements. While longer time horizons might allow for more ambitious goals, shorter frames force the founder to be more concrete. A five-year vision, for example, compels companies to think rigorously about midterm targets. As a company becomes larger and more established, it can set longer-term vision statements, such as Alibaba's twenty-year goals for 2036.
3. **It should project a broad outlook.** Inseparable from your long-term vision is a broader view of the future—both for your venture and its position in the world. Where do you see the industry headed? Will e-commerce become the norm for most businesses, driving job creation moving forward? Vision statements should take a position on where trends are headed and the company's role in that trajectory. Then, founders can work backward to evaluate how to reach that goal. Ultimately, a vision statement that reflects how a company sees the future will help shape the business to meet those demands.
4. **It should be measurable.** Clear metrics are the only way for a company to hold itself accountable. For example, if a firm is running

an e-commerce platform that has the mission of stimulating a region's agricultural sector, one vision statement might be to help fifty thousand farmers within five years use the internet to sell their produce. Specific, quantifiable targets are not just practical but necessary.

5. **It should be altruistic.** This point mirrors an important element of the mission statement: a vision statement should not *only* be about the company. I've often seen vision statements that trumpet the company's wish to be a leader in the industry—for example, the top livestreaming platform or the biggest supplier for online logistics services. None of these are bad goals, but they neglect the customer and society. With greater size and success, a company will likely have more resources to pursue its objectives, but the vision statement should more directly spell out the benefits the company is providing to its customers, employees, partners, and communities.

Alibaba's Vision Journey

It was January 2001, and inside Alibaba's Hangzhou headquarters, Savio led the executive team to a glass whiteboard. Jack had just admitted that, despite a strong implicit understanding of the principles that drove the company forward, nobody had ever codified the key traits in writing. Savio grabbed a marker and popped off the cap.

"Eighty, ten, one," Jack said.

All heads turned to him, puzzled.

Jack said it again. "Eighty, ten, one."

He explained that the first number stood for eighty years—how long he wanted Alibaba to stay in business.

The second number represented the start-up's goal of becoming one of the world's top-ten websites.

And the third number—one—meant that for anyone in the world running a business, Alibaba would be their first and only partner.

This was not bluster. Although it was never written down, Jack had always harbored a bold, really brash vision for Alibaba. In fact, he soon expanded his vision and said he wanted the company to survive 102 years. It occurred to him that since Alibaba was born in 1999, the final year of the twentieth century, a lifespan of 102 years would cover all the twenty-first century and poke into the twenty-second, a company that would have traversed three centuries. It was yet another lofty goal embellished with the type of quirky twist that Jack delighted in.

So how do Jack's vision statements match up to the guidelines?

The first statement offers a specific, long-term time horizon—102 years—but that's basically all it does. Such an expansive time line is certainly ambitious, but it is so vast that it loses specificity, a tangible sense of how to measure progress. That can make it difficult to discern from it the immediate actions that a company needs to take. More importantly, there is nothing linking this vision with Alibaba's mission of making it easy to do business anywhere.

The second statement is also flawed. "Top-ten website" is vague and lacks a clear sense of what exactly is being measured. Is it top ten by traffic, by links out, by revenue generation, or by elegance of web design? The statement also looks inward: How does achieving such a metric benefit anyone but itself? Finally, it suffers from the same problem as the first, bearing no obvious relevance to the company's mission.

The third vision statement may be slightly more redeemable. Being an essential website for business implies providing benefits to Alibaba's main clients, but "essential" is still open to many interpretations without a clear way of measuring progress.

In 2010 we shifted our course and articulated a new vision in response to changes in the business environment. This new vision was in recognition of Jack's view that Alibaba had moved beyond just a platform business and now represented a business ecological system that was to be governed by different rules than a traditional marketplace. Joe Tsai heralded this new vision with an email in 2010 in which he described this new business paradigm.

Previously, we had been positioned chiefly as an e-commerce marketplace—a place on the internet for buyers and sellers to meet and trade. The company had not just grown in business and revenues in the past decade; the platform had also grown far more complex. It was by

2010 a series of interrelated functions to facilitate all the buying and selling, a digital ecosystem.

The disparate array of players—buyers, sellers, suppliers, agents, service providers, financial intermediaries—operated in networks in which they depended on one another to prosper. The more interwoven and dependent these functions, the more the individual actors had to look beyond their own short-term gains and work to support the health of the whole ecosystem.

It was a principle that Alibaba had already operated under for some time. During the depths of the 2008 global financial crisis, we decided to implement a policy called the Mighty Wind promotion, which lowered our rates by up to 40 percent. Even though we expected to lose money, we believed the discounts would provide a necessary boost for struggling Chinese exporters hit by the global meltdown.

Most shareholders would probably have objected to this move if we had asked them because it threatened Alibaba's profit margins. And business school textbooks, in the thrall of Milton Friedman's single-minded dictum that shareholders come first, would have advised against our heavy discounting. Yet, Alibaba believed that even if short-term profits took a hit, the policy would help customers weather the storm and work more closely with Alibaba in the future, thereby benefiting shareholders in the long term. In the end, we not only boosted our customer loyalty but also increased our sales and profits—a proper lesson for all of us that near-term financial metrics are not always the best guide to wise decision-making.

To reflect and promote Alibaba's new interconnected business paradigm, we unveiled a new company vision in 2010 under the same mission. This time we established just a ten-year time line. We pledged in our vision to "promote a New Business Paradigm that embodies Openness, Transparency, Sharing and Responsibility" by 2020. We would achieve our objectives, we said, by doing the following:

- Becoming the first platform of choice for sharing data
- Being the enterprise that has the happiest people
- Lasting at least 102 years

Compared to the earlier version, these three statements moved us closer to embracing a broader, community-focused enterprise, focusing more on the benefit of others rather than simply advancing our own financial interests. They also introduced a concrete, ten-year time period, even if the targets still fell short of providing specific, measurable dimensions.

We continued to adapt to changes in our digital marketplace, and in 2016, we announced another vision statement. All four pillars of the Alibaba ecosystem were well established and running by that time, so we were better positioned to understand how the future would look and what our role would be in it. We provided a short preamble of sorts, concisely describing what was important to us and, crucially, what was not: "We do not pursue size or power; we aspire to be a good company that will last for 102 years."

Following that prelude, the goals were as follows:

- To help ten million SMEs become profitable
- To create one hundred million jobs for the world
- To serve two billion consumers

The key distinction in the first statement was that we wanted to help SMEs become *profitable*—a more specific measure that sharpened the company's aim. It clarified the kind of impact we sought—not just to provide new tools and platforms for SMEs but also to help ensure that they could use them to achieve the best results possible.

The second statement was a continuation of our long-term objectives but again with the emphasis on lifting our communities up in a tangible way, creating jobs. And our third vision statement, far from just casually doubling our 2010 goal—to serve one billion consumers, which might have suggested we remained focused largely on China—made clear that Alibaba would look beyond its home and build a presence across the globe.

Each of these targets held to Alibaba's mission to make it easy to do business anywhere. By empowering SMEs, creating jobs, and serving more consumers around the world, the company was paving the way to a future of more frictionless commerce and inclusive prosperity. That was the model for advancing Alibaba's own success.

Values: The Guideposts Keeping Us on Track

Back in 2001, when we were first confronting the challenge of codifying our mission and values, Savio had scrawled Jack's early goals onto that glass writing board in our headquarters and then turned back. "Next, we need to talk about values," he said.

"Values!" Jack bellowed. "We have lots of values."

He pointed toward his office, and Jin Jianhang, one of Alibaba's co-founders, disappeared inside to search for something. In truth, it looked more like a warehouse than an office. Files and documents were strewn in imposing stacks, some towering above the desk. But within the clutter was just what they needed. Jianhang returned with a ream of papers, at least fifty pages worth of notes on the ideas and business concepts Jack had uttered as he built Alibaba.

Savio gulped at the size of the stack. Recruiting Lucy Peng into the effort, he divided the notes into two piles. And with that, they went to work trying to learn from what had been a haphazard array of observations and statements of principle.

> Alibaba is a unique company because of our culture and values. It is our values that make us who we are and set us apart. Our values enable us to make the right decisions at the most critical moments.
>
> —Jack Ma, founder of Alibaba

A company's values are not the same as an individual's personal values. Personal values touch on morality, on right and wrong. When we talk about the values of an organization, we are referring to the codes of conduct that direct how its members behave and make decisions, the objectives that give the organization meaning. These often take the form of distinct rules and policies, but they also refer to a less tangible mindset around the company's internal collaborations and commitments.

As Alibaba's founder, Jack led primarily by example, confident in his awareness of who *he* was. Imagine if an entrepreneur simply "borrowed" the values that he or she came across from, say, the website of another firm. That entrepreneur might end up with a list that sounded good on paper, but it would lack meaning and would not serve as a true guide to behavior and practices. Every company needs to articulate values true to its own nature.

> Our values truly resonate with us. Our values are our unique culture which powerfully supports us as we continue into the future.
> —Daniel Zhang, CEO of Alibaba Group

Every founder must carefully choose the values that resonate first with him- or herself *before* those values can resonate deeply with the company. This is also the reason why founders should never fully delegate the hiring process. Especially for smaller companies with limited head count, the person in charge should always try to meet with every new candidate. Assessing whether potential employees are the right fit means assessing, among other things, whether they exemplify the proper values for that particular enterprise. One way to help turn values into a culture is to choose employees who share them.

> Everyone has values. Alibaba's values are the key for [our employees] to get along with one another. They are our shared commitment to who we are and how we do things.
> —Lucy Peng, Alibaba group partner and former CPO, and Ant Group former CEO and chairwoman

Every employee needs to determine if he or she identifies with a company's values, especially in the start-up phase. A shared understanding of a start-up's values is crucial to helping everyone weather the inevitable difficult times. Teams bound by a common commitment work more effectively under stress, recover more quickly from setbacks, and seize on opportunities with greater focus and energy.

Values are like traffic signs, ultimately, laying out clear guidelines for everyone. Though red lights and speed limits can seem restrictive, their purpose is the common good, security, and confidence that everyone is adhering to a shared set of principles.

In all my travels, the city with seemingly the most disorderly traffic, hands down, is Ho Chi Minh City in Vietnam. If, as a pedestrian, you wait for traffic to stop or a road to clear before crossing the street, good luck, you'll never budge. Instead, pedestrians, in a terrifying, remarkably adroit dance, simply wade into the vehicular chaos—a sea of motor scooters, bicycles, cars, and trucks.

The first time I saw this, I expected the worst and waited for what I was certain would be a bloody catastrophe. Little did I know, until observing this for a while, that an invisible system was at work, guided by unwritten but shared values. Drivers adjust their speed and direction as necessary to safely allow pedestrians through and to navigate their vehicles to their destinations. It has become second nature. They are attentive without seeming to be. Although it is alarming to first-time visitors, eventually they make it across the streets without incident—or at least I hope they do.

An enterprise's values are like traffic rules, whether explicit or implicit: they guide employees to meet their own objectives with confidence and contribute to the smooth functioning of the enterprise as a whole.

Back to our project of recording Alibaba's values: Savio and Lucy took notes from the pages before them and jotted things down on the glass whiteboard. Eventually, Jack's ideas—principles, inspirations, and musings—covered the whole wall in a muddle of English and Chinese characters. The Alibaba executives surveyed the bits and pieces and then began to discuss how they could be woven into a clear narrative.

This point relates to that one.

Those notes can be combined.

That policy is outdated. Toss it.

Daylight gave way to nightfall, and still the discussion flowed. Seven hours had finally passed when they paused to take stock. The ideas had been scrutinized and evaluated many times over. What remained were nine values that defined Alibaba's culture.

From Nine Invincible Swords to Six Vein Spirit Swords

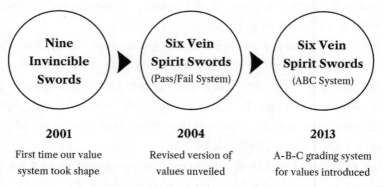

The evolution of Alibaba values.

The Alibaba executives needed a way to bring the ideas to life, and early in their conversations a good way of doing that surfaced. Savio had been surprised to discover that he and Jack were both childhood devotees of the popular Chinese author Jin Yong. Jin Yong is well known for his novels that are steeped in *wuxia* (武侠)—or martial arts—and the chivalrous culture embraced by legendary masters. He spun fantastic tales of glorious, humble fighters, rooted in Chinese history. The stories were about great feats of heroism and sacrifice. Savio and Jack quickly bonded over their shared passion for these gallant warriors, who were always battling on behalf of the oppressed. Rather than besting their opponents with strength, Jin Yong's heroes relied on cunning and ingenuity.

Inspired by that chivalrous world, Jack and the executive team decided upon the name Dugu Jiujian (独孤九剑) or Nine Invincible Swords to organize and frame the company's values. The idea was that once these "swords" were mastered then, like the protagonists in Jin Yong's novels, our employees would have the strength and fortitude to confront any resistance.

In 2004, Alibaba was pouring its efforts into making Taobao China's leading consumer marketplace. Waves of new employees were hired, creating a need to streamline the company's values into a more concise set that these new employees could easily grasp. That year, Savio orchestrated a focus group of three hundred employees, and with their input, he condensed the original nine values into six, which were called Liumai Shenjian (六脉神剑) or the Six Vein Spirit Swords, named after the heroic sword skills of a famous *wuxia* character, Duan Yu, from Jin Yong's novel *Demi-Gods and Semi-Devils*.

The Six Vein Spirit Swords that were agreed on were customer first, teamwork, embrace change, integrity, passion, and commitment. For the company's employees, these "swords" served as both weapon and shield, empowering them to achieve great things on behalf of the little guys, the Chinese SMEs previously shut out from commercial prosperity. Far from being mere buzzwords, these values were upheld with seriousness at all levels of the company.

A New Set of Values

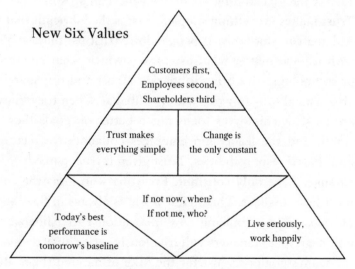

New Six Values

Customers first,
Employees second,
Shareholders third

Trust makes
everything simple

Change is
the only constant

If not now, when?
If not me, who?

Today's best
performance is
tomorrow's baseline

Live seriously,
work happily

Alibaba's most recent six values reflect the changing times of the company and its markets.

After fifteen years following that set of tenets, Jack and the leadership team later conjured up a new set to reflect the development of Alibaba's ecosystem model and the changing digital economy. Alibaba unveiled its New Six Values in 2019, a set of principles not just for the company but to also fit the new era we had entered.

Our emphasis was now very clearly placed on supporting inclusive growth, spreading prosperity widely, empowering entrepreneurs, collaborating, and partnering with others—not obsessing on enriching our shareholders alone.

1. **Customers first, employees second, shareholders third.** This formally enshrines in the Alibaba value system Jack's conviction that our customers, fulfilling their needs, are the whole reason our business exists. Employees come second because they make up the teams that devise products, serve our customers, and ensure customer satisfaction. They are the ones who create value—not the shareholders, who play an important role but come third. If the employees are working hard, then the customers will be happy, and the increased value will naturally benefit the shareholders. That is the logical progression we believe in at Alibaba.

2. **Trust makes everything simple.** Trust is the lubricant that makes Alibaba run smoothly. If people don't trust us, they won't work with us—no matter what assets we own or what new products or technologies we bring to market. Trust and transparency are also critical to teamwork within Alibaba. When they know they are trusted, employees focus much better on goals they understand, and the leadership embraces collaborative relationships and shared responsibilities. Trust given is trust gained.

3. **Change is the only constant.** Everyone who joins the company must face this fact. The digital world is always in flux, shedding old ideas and embracing new innovations. At Alibaba, it does not mean that managers will frivolously revise objectives or move the team's goalposts just for the sake of moving them. But as a globally competitive technology firm providing a full ecosystem of business services, all employees must be able to adapt to the industry's breakneck pace and changing dynamics.

4. **Today's best performance is tomorrow's baseline.** This value encourages constant improvement. Competitors are constantly closing in, while customer expectations always change as customers grow to expect more and better services. At Alibaba, employees can never get complacent.

5. **If not now, when? If not me, who?** This famous saying was used in Alibaba's very first job listing in Hangzhou-based *Qianjiang Evening News* in 1999 to attract new employees for a budding new venture. This value is intended to encourage every employee to believe in his or her own contributions and to seek to make a difference. At Alibaba, every employee is integral to the company; the employees are all uniquely suited to help serve the company's greater mission, solve problems, and innovate.

6. **Live seriously, work happily.** Alibaba's final value flips the more commonly held belief—to work seriously and live happily—on its head. The unorthodox formulation is inspired by the belief that our employees should want to "live deliberately," as Henry David Thoreau put it in his famous book, *Walden*. We hope that they ask themselves what they want most from life, what has meaning for them, what principles are most valuable in helping them navigate challenges—in other words, living seriously. Once they have answered these questions, they can decide what sort of career choice will fulfill their personal objectives. If life and work are aligned, work then naturally becomes more enjoyable because employees are finding and enhancing what is meaningful for them.

A Hierarchy of Values

These values, while all important, exist in a hierarchy. They operate on different levels: the value judgment level, the organizational level, and the individual level.

At the top of the pyramid is the value judgment level. "Customers first, employees second, shareholders third" is Alibaba's highest order value; it is the firm's top priority and informs decision-making. The middle segment represents the organizational level implemented by management. Building trust and embracing change are values that should permeate interactions throughout the entire organization, as employees strive to trust one another and hold themselves to that standard. In the final segment, the individual-level values offer guidance for each employee.

When a company is clear about its values, they act as a filter to help identify the kinds of people who will prosper and thrive in the company's

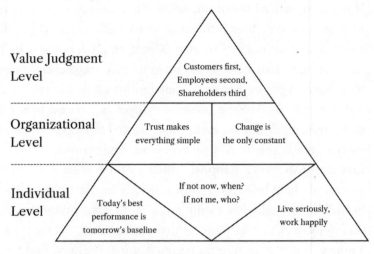

Value Judgment Level

Customers first, Employees second, Shareholders third

Organizational Level

Trust makes everything simple

Change is the only constant

Individual Level

Today's best performance is tomorrow's baseline

If not now, when? If not me, who?

Live seriously, work happily

Hierarchy of values.

work environment and contribute the most to success. Internally, they can also become part of a navigation system for setting priorities, reducing friction, and overcoming obstacles.

As a company grows, values become even more important, enabling employees to make the right decisions in what can be a stressful environment, without needing constant oversight. At the same time, these principles should empower junior staff to participate in discussions around team objectives and even push back on ideas that may steer the team away from core values.

In the end, a company's value system carries the recognized set of beliefs for everyone to work together, collaborate, and innovate. Values unify everyone under a common banner.

Communicating the Message

The potency of a strong value set is wasted if it is not effectively communicated. To introduce and socialize Alibaba's updated values to our thousands of staff across the world, we launched the New Six Values international tour in October 2019.

This was not quite a Rolling Stones world tour, but we devoted months to intensive planning of all the details to ensure our values were both understood and embraced. We held dozens of meetings on how the new values should be articulated and how each of us on the committee understood their meaning through the prism of our own experiences.

Regional representatives from the United States, Europe, Southeast Asia, and India were all present to provide a wide range of perspectives. Once our new value set was finally ready to be unveiled, the company's executive team in Hangzhou divided territories and hit the road, flying to different overseas markets and teaming up with the local leadership to conduct workshops.

I accompanied Jane Jiang, Alibaba's international chief people officer and one of the company cofounders, to Paris to convene with department heads and their direct reports from our London, Milan, Munich, Madrid, and Amsterdam offices. A key aim was personalizing the New Six Values, sharing our own interpretations, so that the workshops would involve lots of useful dialogue.

When we discussed "Trust makes everything simple," for example, I spoke at length about the camaraderie I felt working "in the trenches" alongside my Hangzhou colleagues, the sense of a common cause we shared. I explained that, twenty years later, I still felt that bond. I also shared stories about meetings with contractors, suppliers, and collaborators, who often said that the dedication and passion of Alibaba employees was unmatched.

Altogether, the sessions were eye-opening for us, a two-way interaction. We had arrived to share with our European colleagues Alibaba's updated company values, but I think Jane and I ended up learning just as much from hearing the views of the employees at the gatherings. Our workshops became a space where people could candidly share their own thoughts—not just about the practices that would or wouldn't work in their regional offices but also their personal beliefs and philosophies about the meaning of work and how to approach it.

Listening to their ideas, I was struck by the real ambition of what we were even attempting to do: we were respecting the rich texture and cultural diversity of the European markets, while also introducing the best of Chinese work culture. It was a daunting but also invigorating effort to strengthen the complex social dimension of Alibaba's approach to the digital economy.

Mission, Vision, and Values and the Tao of Alibaba

The development path of a company is not unlike that of a group of individuals going on a journey to a common destination.

During a training session I led for entrepreneurs from Southeast Asia, a student from Vietnam contributed this wonderful illustration as a way to visualize the core concepts of mission, vision, and values. In the sketch, a trio of travelers embark down a long path that stretches toward a mountain in the distance. But no two journeys are ever exactly the same. Every person will interpret the twists and turns, the challenges, the achievements differently, based on his or her background and circumstances. The drawing is about life and direction, not just business and management. As an entrepreneur myself, the image resonates strongly.

The horizon in the distance with the sun above represents our mission. Our mission is the big picture in the far distance. It establishes a direction, but it is never quite attainable. Yet it offers us clarity and lights our path forward, even when the road is steep.

Our vision is represented by the peak. It is distant, but that vision is what pulls us forward. To fully articulate the scope of our vision, we might conjure even more measurable goals, like, say, to scale five of the seven summits in the next three years to give us greater motivation.

And finally, our values are symbolized by the oversized compass in the drawing, wedged curiously into the ground. The compass not only provides direction; it also gives us a coherent system of guidelines that will help the three travelers stay grounded and on course by banding together as a unit to reach the destination.

But there remains one element not quite fully captured in the visual rendering of these concepts worth mentioning—the spirit with which you embark on your journey. Are you dragging your feet, complaining about the heat and the weight of your backpack? Or is there a spring in your step as you stride toward the destination? Are you trying to summit a mountain for a one-off thrill, or are you embracing this challenge as central to your life journey? How does your mission and purpose influence the way you proceed along and see your path?

When I was a kid growing up in Palo Alto, I took martial arts lessons from a master who often said, "Life is kung fu, and kung fu is life." I was young—absorbed by how I looked in all those cool stances and graceful maneuvers—so I didn't think about his words much at the time. But years later, after joining Alibaba, I would often hear Jack say, "Life is Alibaba, and Alibaba is life." The resonance was striking. I was now hearing that message in a new way.

It would take years of observing Jack and working at Alibaba before I started to make sense of this simple piece of wisdom, and years more before I started to inhabit the mind-set it suggested. What Jack meant was that by becoming a part of Alibaba, you were committing to more than just a job: you were choosing to embrace a mission, to hold yourself against a higher set of standards, and committing yourself to having an impact beyond the nine-to-five demands of the job. This is the essence of the tao that permeated our work and life, connecting the two. And it is a topic we will continue to address.

Just like athletes with their chosen sport, or painters with their art, successful entrepreneurs must have an inner purpose, erasing many of the boundaries between life and work. That, I have found, is where great achievements begin.

6

Strategy

The Road Map

There were moments when working at Alibaba felt like jumping into an inviting blue ocean. The horizon glowed with promise and the shimmering expanse appeared endless. There were cool onshore winds, waves of opportunity, one after the other. And working under a leader like Jack, who empowered his young team to lead major projects and make key decisions, only amplified our desire to dive right in.

In only my first year at the company, Jack appointed me chief producer of our international English-language website. My team brainstormed new ways to serve our international customers, such as useful online sourcing tools to make it easier to access trade data and new interfaces to help international buyers use Alibaba.com's product search functions. Our proposals were usually quickly approved and rapidly implemented.

But inevitably that wide blue ocean grew choppy. We loved our operational freedom, but at times it felt like too much of a good thing. The opportunities started to overwhelm us; we tried to do too much at once, pushing a lean team toward exhaustion and burnout. On several occasions, after rushing out a new feature, we would find out that it was shoddily designed or plagued with bugs, and yet our manpower was stretched too thin to properly address the issues. For example, at one point in the early days of Alibaba.com, we tried to design a "business center" concept to provide SMEs all the trade services they might need, such as shipping costs, product inspection services, currency conversion tools, general trade news, commodity prices, and more. This product ended up being a hodgepodge of services lacking focus and void of a compelling value proposition. Experience would eventually show what we lacked in those early years: a clear strategy for execution.

Why Strategy Matters

Strategy is the unifying component of the Tao of Alibaba.

A company's strategy is critical to achieving sustained success. In our Tao of Alibaba framework, mission, vision, and values (MVV) provide overall the direction and spirit for the organization. Strategy is positioned directly below it, as the centerpiece of the pyramid, offering the road map or principles (or *li* [理] as referred to in Chapter 4) the company will follow to serve its mission.

If MVV is the *why* and the *what*, then strategy describes the *how*. It treats the vision as a starting point and then maps out the difficult decisions and actions required to get there. It also connects MVV to the two essential resources in a company's arsenal—organization and people and performance management—ensuring that the tools and tactics employed are all aligned with a company's purpose and values.

While Alibaba has taken many missteps in its journey, the company learned how to mold these hard-won lessons into valuable frameworks, which enabled Alibaba to navigate the remarkably difficult, always changing digital business landscape with greater confidence.

That did not come easily, but over the years Alibaba's strategy solidified. We called it B2B2C, meaning business-to-business-to-consumer. To fulfill the company's mission of making it easy to do business anywhere, we had to cut out the middle players and create a full-service value chain that links manufacturers and suppliers (the first B) with the distributors and retailers (the second B) with the ultimate consumers (C).

B2B2C

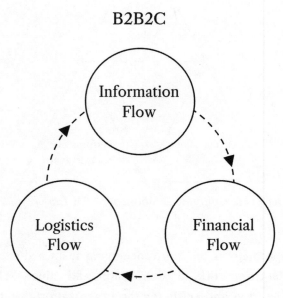

The three elements of the flow of commerce.

Later, we realized that covering this whole sequence end to end required arranging three key elements, which we described as flows:

1. Information flow: connecting buyers and sellers through Alibaba's original marketplace platforms.
2. Financial flow: ensuring that these merchants and clients could safely and reliably conduct transactions with one another.
3. Logistics flow: providing the capability to move products from point A to point B on the desired schedule.

Seeing the sequence as three distinct flows provided a clarifying framework for how we needed to proceed. Even so, there were many gaps that we had to fill as we went along. These included what order we should follow for implementation or when to address the pain points that arose at each stage. Having the larger structure in place, however, gave us the confidence that we were moving in the right direction.

B2B2C, along with other important organizational concepts, was developed by Professor Ming Zeng, Alibaba's former chief strategy officer. Zeng, a PhD in international business, met Jack for the first time in 2000 when conducting research on Chinese management practices. The two kept in touch. In 2002, Zeng was one of seven founding faculty at China's first private business school and, at Jack's invitation, began regularly helping facilitate high-level strategy discussions at Alibaba. In 2006, he joined the company full-time.

Zeng likes to talk about strategy as a combination of science, art, and craftsmanship. Business schools and management experts have long tried to conceive universal laws of strategy from painstaking analysis of business performance and case studies. These theories and principles represent the science, and they can be useful—up to a point. The best entrepreneurs often exhibit a touch of creative genius—whether a maverick flair or the intuitive ability to imagine and inspire—that cannot be taught in any classroom. That is the art.

And finally, there is the craft, defined by Zeng as the "aspects of strategic thinking that simply rely on accumulated knowledge and wisdom" honed through patience and practice. We will focus on the proven steps of craft, which every company can readily benefit from, while shedding some light on the scientific and creative elements as well.

This chapter also explores the three rings of strategy making—*want*, *may*, and *can*—and describes how their points of intersection define a company's final strategy. Alibaba has come a long way since my chaotic early days there, and I saw how thoughtful strategy statements helped us cut through the noise and amplify our strengths. I also observed how the steps of strategy mapping, the most important factors in executing a successful strategy, functioned in Alibaba's own trials and triumphs.

The Three Rings of Strategy

Strategy lies at the heart of business management, determining a company's positioning, priorities, and lines of focus. But how do we *formulate* these strategies?

Strategic thinking begins with a set of three considerations, known within Alibaba as the *three rings of strategy*:

1. What you *want* to do—your dreams.
2. What you *may* do—your opportunities.
3. What you *can* do—your capabilities.

The Three Rings of Strategy

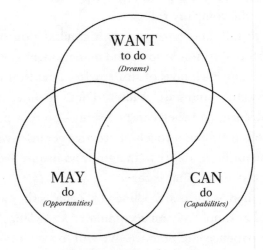

The three rings of strategy.

The first consideration—what you *want* to do—is closely linked with your company's mission and vision. It's the pie in the sky, the dream that occupies your imagination. It's the passion that excites you and the highest specific aspiration of your strategy.

The second consideration—what you *may* do—looks outward. It involves analyzing the external conditions of the market and determining the circumstances you and your business must navigate. What are the trends? Where are the opportunities? Which products and services are heating up, and which have already peaked and started their decline? Within tech and the digital economy in particular, the buzziest start-ups are frequently just passing fads, shining bright for a few months or years before burning out into obscurity.

The final consideration asks you to assess pragmatically what you *can* do. This requires facing the reality of your capabilities. What are your strengths? Your resources? In which areas are you or your team most capable or experienced?

In facing these questions, if they do so honestly, entrepreneurs often find a wide gap between what they *want* to do and what they *can* do. This can be a painful pill to swallow, but avoiding it is even worse. Rather than basing plans on an illusion, acknowledging any such gaps between aspirations and capabilities early on is the only way to ensure that subsequent strategies can succeed.

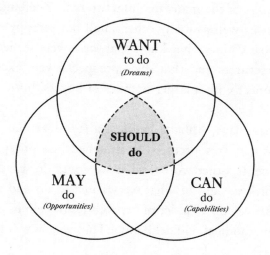

SHOULD do = STRATEGY

Together, the three rings make strategy.

Methodically working through each of these considerations helps inform the really critical issue: what a firm *should* do. In the diagram, the central field where the three rings overlap represents what should be done. This is the magic spot where an animating dream finds a realistic path to taking flight.

When Jack launched Alibaba in 1999, the pie he dreamed of was not so much in the sky as making up the entire sky. He aspired to build a global digital company that could elevate the entire B2B marketplace, outshining even the most successful, well-established dot-com businesses. At the time, the stars of B2B e-commerce were Commerce One and Ariba. Though both firms would later tumble following the dot-com crash, they loomed at that time as imposing competitors. They forced us to confront hard truths about our assets and limitations.

The reality was that our technology couldn't compare with the cutting-edge firms in Silicon Valley. We didn't have as much venture funding, and our clients weren't as sophisticated. Even when it came to positioning the company as a kind of online trade show, we did not have as much trade show experience or credibility as, for example, Global Sources, another competitor.

What Alibaba did have, however, was an unmatched understanding of the Chinese market. In particular, Jack and his cofounders possessed a deep knowledge of the many manufacturers in Zhejiang and Guangdong provinces as well as many other small but scrappy entrepreneurs spread out across China. They knew that these were skilled, highly motivated manufacturers and that their prospects were excellent if only they could connect with a larger number of buyers, both within China and abroad.

In those early days, Alibaba had just a few dozen employees and a simple platform. Its resources were stretched just providing a simple digital interface. In other words, what we could do was quite limited, but we did not lose sight of what we wanted to do and saw ourselves as working our way up a ladder of capabilities to help us get there. Our actions were methodical and deliberate. That was how it began.

Want, May, Can: The Taobao Story

A crucial early phase in Taobao's trajectory provides a more detailed case study of how these three rings help conceptualize a successful strategy. This goes back to 2005, when Taobao, our early consumer platform, was less than two years old and dreaming big. Driven by Alibaba's mission to make it easy to do business anywhere, what Taobao *wanted* to do was become the biggest consumer shopping website in the world.

As for what it *may* do, several options were being debated. First was to create a B2C marketplace, helping brands get online and sell directly to customers. Another option was to mimic eBay's model and build out a C2C platform, focusing on getting Chinese individuals and microbusinesses selling online. Finally, there was an emerging category of shopping search engines—web services that enabled users to compare products and prices from multiple sellers to find the best buy.

Taobao managers also considered producing a version of the online advertising platforms that were gaining traction, including the classifieds model popularized by Craigslist. Before plotting a course of action, the managers decided to undertake a deep analysis of the field. What was *actually* happening in the market? Which options *could* Taobao feasibly handle?

Many of the Taobao staff were young recent university graduates, full of energy and passion, and the venture was still very much a start-up. What the staff lacked in the established pedigrees, technical skills, or global outlook of their peers in Silicon Valley they made up for with strong local knowledge of Chinese consumers and an intense desire to make an impact. Taobao also benefited from Alibaba's highly successful B2B platforms. They provided operational expertise and could furnish many replicable best practices.

By laying out all the various considerations of Taobao's want, may, and can considerations, a clearer, comprehensive picture of how it *should* proceed came into focus. Given Taobao's core strength of localized Chinese market knowledge and operating teams, the team decided to forego the international markets and focus on serving Chinese consumers. This decision leveraged Alibaba's competitive strengths and mitigated its

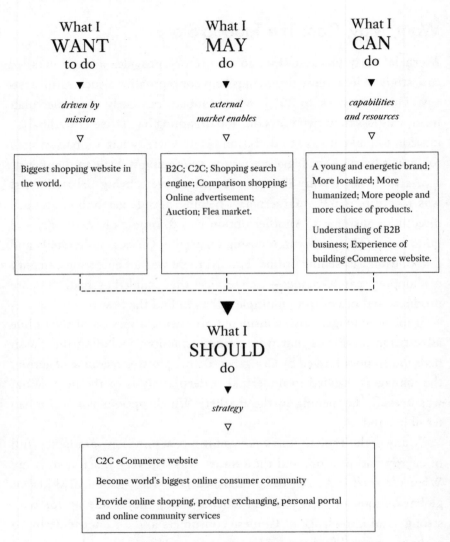

The intersection of Taobao's three rings form a strategy.

shortcomings. As a new platform, the team also recognized the challenges of convincing sophisticated brands to offer their products on the rudimentary Taobao platform and so opted instead for the C2C approach.

Crucially, they also adjusted their ambitions. Rather than striving to become the biggest shopping *website* in the world, they understood that the internet was still very new and unfamiliar to many Chinese. To address their concerns about online shopping, the team determined that it had to instill a human element in the Taobao platform, to open it to conversations that would provide more information. So they allowed people to post questions, share comments, and openly discuss their experiences. This reflected a subtle but important adjustment in the team's aim: now, they were offering customers the world's biggest online shopping *community*.

This strategy allowed Taobao to increase its value to its customers beyond just online shopping. Taobao would not simply be a place to conduct transactions but an inviting and navigable environment for Chinese users to interact with one another, discuss new products, and fulfill their needs.

From "Should Do" to Strategy Statement

Defining *want*, *may*, and *can* shapes the understanding of what a firm *should* do. This then forms the basis for a company's *strategy statement*—an indispensable credo that will drive important decisions at every level of an organization.

Should do ► strategy statement

Based on your "Should do," you may design your
strategy statement using the WHO, WHAT and HOW format.

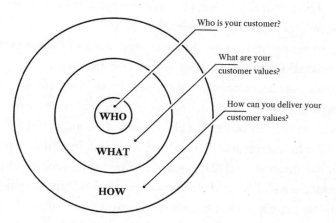

Who is your customer?

What are your
customer values?

How can you deliver your
customer values?

WHO

WHAT

HOW

*Who, what, and how are the key components
for any strategy statement.*

At Alibaba, a cogent strategy statement answers three essential questions:

Who is your customer?
What are your customer values?
How can you deliver your customer values?

Taobao, again, presents us with an instructive model to demonstrate this process in practice.

Step One: Who?

When Jack began putting a few trusted commanders in place to start building Taobao in 2003—from "zero to one," as we jokingly said—the new business had to devise a strategy statement that could chart a path through the competitive market.

The first question—Who is your customer?—might strike many as obvious. Yet it's important to think *past* the obvious here, looking at the nuances of the different types of customers, their different conditions and different values. eBay, arguably the world's most powerful tech company in those days, had already announced its plan to sweep up the Chinese market and had promptly seized what it said was a 95 percent market share, so Taobao needed great clarity if it was to have any hope of breaking this stranglehold.

That imposing figure—95 percent—appeared to be an impregnable wall. How could Alibaba possibly compete with that?

Taobao's managers considered the data and decided the prevailing narrative might have misconstrued the market reality. It turned out that the 95 percent market share that eBay boasted about referred to the proportion of *online* consumers, meaning everyone in China who (1) had access to the internet and (2) was already spending money through online transactions. This total "addressable market" amounted to about three million people at the time, of which eBay had captured a formidable 95 percent. But in a country of more than a billion people, that was a very narrow segment of the population.

Taobao then did a deeper examination of its potential customers. Was it really just going to have to struggle to win over the 5 percent of online consumers who weren't yet using eBay? The clear answer was that there was a different customer base out there to be pursued—a far bigger one. Taobao chose to take aim at the broader pool of Chinese netizens—people connected to the web, browsing and chatting, but who had not yet embraced online shopping. This group included roughly ninety million people, our research showed, effectively multiplying eBay's total addressable market thirty times over. Many of those people were, we believed, ready to try e-commerce.

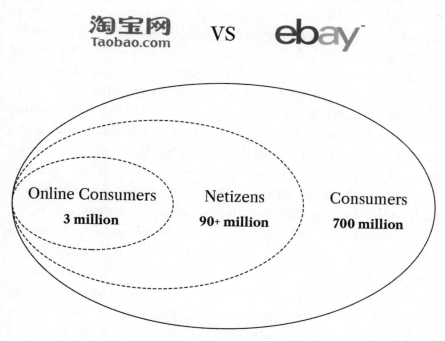

Determining who your customer is can be more complex than it may seem.

It didn't end there. Taobao's true sights were set on the full seven hundred million consumers in China—people underserved by traditional retail options but who had not yet been pulled online. By widening the aperture through which it viewed its customer base—by thinking *past* the obvious—Taobao discerned a vastly greater opportunity than what others in the market had identified.

Step Two: What?

After settling on *who* its customers are, a firm must determine *what* it stands to offer them. These are customer values—the characteristics a firm embodies to draw people to it and to serve them.

There are two complementary aspects to customer values: consumer needs and a company's offerings to its customers. One of the best ways to deepen the understanding of consumer needs is to look at times when they are not served well, when there is friction. Questions to ask are if you can resolve those pain points, and if the needs are truly essential to the customers or just things that would be nice to have. Does your company offer unique value, or can competitors claim to offer similar benefits? Will your customers recognize your unique values?

When I was first living in China as a student in the nineties, buying anything was an ordeal. Shopping meant spending at least half a day at various stores, trying to track items down. Sales help was mostly nonexistent. Choice was extremely limited, and comparison shopping all but impossible.

Taobao delivered a significantly better experience when it launched. It offered a far greater selection, all from the convenience of your home, on a computer (and later, of course, your mobile phone). Vendors in the online marketplace were incentivized to lower their prices to lure customers from competitors. And customers could feel more secure in their purchases because Taobao's website featured a very active section for feedback and reviews. By sharing their experiences on the products they'd purchased, shoppers boosted the confidence of other netizens interested in using the platform.

Making shopping safer, cheaper, and easier was exactly what Taobao had set out to achieve. These three characteristics belong to a category we call *survival* customer values, values that were unaddressed by others in the market.

Beyond survival, there are *thrive* customer values, which focus on elevating users to better enjoy their shopping experiences. Taobao users could create profiles, post questions and comments, and interact with others as part of a vibrant grassroots community. Eventually, some users built reputations for being knowledgeable about certain kinds of products and developed a following, while others spurred engagement through recommendations—what today we would call social media influencers. It was a vital and lasting lesson for Alibaba, as well as its customers, that e-commerce could be *fun*.

Taobao's responses to the values of survive and thrive differentiated it from its competitors, which is exactly what gave it an edge over eBay. Considering the survive and thrive elements together makes the *what* part of any company's strategy statement as strong as possible.

Step Three: How?

The last element of a strategy statement should address *how* an organization intends to deliver these values to their customers.

Because Taobao planned to seek customers who were not already shopping online, its outreach campaigns were not limited to the web. Television commercials, particularly targeted at young people, were used to introduce people to the C2C marketplace. Crucially, we also organized offline meetups, where groups of young people could learn about all the things the website empowered them to do. Most internet firms ignored this offline population, but we knew this was a big pool of prospects who could significantly expand the overall user base. The aim was to provide such a reliable, engaging, and all-encompassing experience that shoppers would no longer feel they needed to go anywhere else.

These are all components of a successful strategy statement. At this point, much of the hard work is done and the actual statement follows a simple formula:

Your company's strategy is to provide [what] *to* [who] *by* [how].

With Taobao, we ended up with:

Taobao's strategy is to **provide** *a safer, cheaper, more convenient, low-barrier, entertaining, and fun experience* **to** *Chinese consumers* **by** *creating the largest online consumer platform and community.*

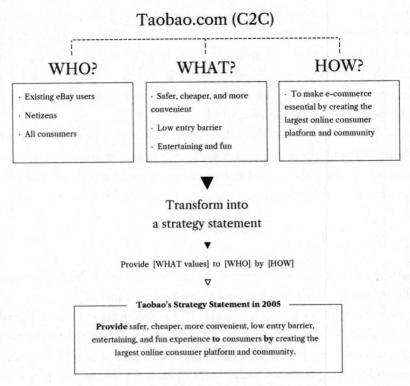

The who, what, and why of Taobao.

In the previous chapter, I discussed Disney's revised mission statement. It states: "The mission of The Walt Disney Company is to entertain, inform and inspire people around the globe through the power of unparalleled storytelling, reflecting the iconic brands, creative minds and innovative technologies that make ours the world's premier entertainment company."

To me, this reads much more like what we at Alibaba would designate as a strategy, not a mission statement. Your company's mission and vision should be direct, uplifting, and easy to remember, while the strategy statement should explain how, exactly, the firm will get things done.

The Key Success Factors

It's important to remember that a strategy statement is not just a label. Its true contribution is in clarifying a company's intentions and affirming the viability of its chosen path. It's paramount that this path aligns with the company's mission, vision, and values. There should be harmony between mission and strategy statements, complementary voices in a choir.

(1) Define a clear outlook for the future.

"The company that sees the present *from* the future has strategy. The company that simply moves from the present *to* the future has no strategy," says Professor Zeng.

In other words, it's important to have an outlook—a clear sense of how you expect the market landscape to evolve and how you will respond to it. Arriving at such a perspective shapes a company's position and illuminates the key points of a successful strategy. Or, as Zeng likes to say, "Start with the end and work in reverse."

That was the case in 2006, when Taobao was winning its competition with eBay. I wasn't working directly on Taobao that year—I was senior director of Alibaba's global business development—but I watched intently from the sidelines as Taobao surpassed the Silicon Valley darling. Taobao hit a milestone when eBay decided to withdraw from China, affirming its strategic direction and execution. Still, team members knew it was no time to rest on their laurels. They turned their thoughts to what would come next.

The graphs on the next page come from actual slides used during the company's strategy meetings in 2006. As the first graph shows, the Taobao team was already forecasting what the market landscape would look like six years out. The pie chart presents a more detailed breakdown of their projections.

The largest slice of the pie belongs to "Online Shopping Mall," which represented 40 percent of China's estimated 1.4 trillion RMB e-commerce market. Armed with this analysis, Taobao leadership decided to initiate a new push into the B2C segment. They soon began building the Taobao Mall website to pursue this new opportunity. Launched in 2008, Tmall, as it is known today, was Alibaba's means of pursuing that market opening.

As the Taobao team knew from the earlier strategic conversations, getting well-known brands on the platform was key, but they initially found that a struggle. E-commerce in general was still fighting to gain

Strategic Thinking ▶ Future Outlook

China's Consumer-oriented market has huge potential

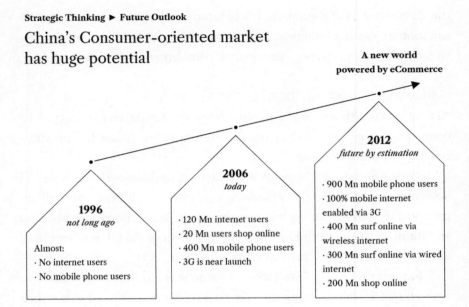

Strategic Thinking ▶ Sector Comparison

Every sector is huge in size

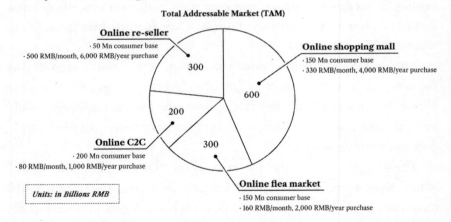

Taobao's future outlook, as of 2006.

acceptance with Chinese consumers, and many prominent brands worried that the association with Taobao, which had developed a reputation as something of an online flea market due to its focus on smaller, mom-and-pop sellers, would dilute their images.

Using my contacts in global business development, I set up a partnership meeting with Procter & Gamble. We had started working with the company on a plan to support its procurement operations via our B2B platform, but we also tried to pitch P&G the idea of selling its products directly to Chinese consumers on our platform. Their senior executives were skeptical, but we managed to strike a compromise; we persuaded them to test the market by offering a single product line on Tmall—their Gillette razors. Most other big brands were not even willing to go that far. But it was a start.

Ironically, the 2008 global financial crisis actually helped us. The big multinational consumer brands, which sourced lots of products and materials from Chinese producers through Alibaba platforms, experienced collapsing demand in the West as economies slid into recession. Many Chinese producers responded by focusing on China itself for a change. They developed their own local brands and, in some instances, boosted sales by turning to Tmall.

By the time the financial crisis subsided and economies started to recover, those multinational brands observed the successful growth of the Tmall platform and started to embrace it as a viable sales channel. Today, Tmall's gross merchandise volume is the largest contributor to Alibaba's retail business performance. And the Gillette razor experiment flourished. Since that initial test, Proctor & Gamble's sales on Tmall grew a thousand times in ten years, comprising nearly 30 percent of the company's overall sales in China.

(2) Avoid items on your list of not-to-dos.

Good strategy is as much about what you *don't do* as it is about what you actually do. One of the toughest challenges facing any entrepreneur and young company is learning how to say no. When everything looks like a good opportunity, it can feel impossible to choose. But pursuing every lead will often stretch employees too thin, often at the expense of a larger goal. A well-delineated strategic statement helps to keep a start-up focused.

In the early 2000s, when Alibaba was still relatively new and Taobao had just been launched, a number of other internet businesses were starting to take off in China. At the forefront were online video games, which provided an attractive business model. If you produced a hit game, the costs of replicating it were negligible while the profit potential was sky-high.

Alibaba's chosen lane was e-commerce. We thought that, if we wanted to pursue the games business, Alibaba should probably spin off a new division and articulate a new mission for it, something like, "creating happiness through gaming." But then it would no longer be Alibaba. Alibaba was in the business of e-commerce, and that was where we felt comfortable and should focus our resources.

Another hot sector at the time was communications. Instant messaging platforms like Tencent's QQ and MSN Messenger were quickly becoming the preferred way for many Chinese web users to chat. Should we develop a platform of our own? we wondered.

Unlike gaming, enhancing communication *is* an integral part of facilitating business, and Alibaba *did* decide to launch its own messaging tool for Taobao. We called it Aliwangwang. The strategic point, we decided, was not just catching a new trend but also allowing our core customers—buyers and sellers—easier access to one another, increasing the platform's overall level of trust and transparency. The ability to message counterparts directly empowered all parties on the marketplace. This new product, then, aligned with Alibaba's mission and strategy: it *did* make it easier to do business anywhere, and it undoubtedly helped Taobao grow its thriving online community.

(3) Develop sustainability relative to the competition.

It's good strategy to create barriers of entry in a market or protective moats around a business to ward off competitors. At Alibaba, we employed two methods: a first-mover advantage and high switching costs.

Knowing that survival rested on building a critical mass of suppliers and buyers, one of Alibaba's early priorities was to attract new users. So in our early days the team launched an aggressive outreach campaign to get the word out to SMEs. Staffers contacted business associations, attended trade shows, pored over catalogs and databases for any leads they could find. By being one of the first such websites out of the gate, Alibaba managed to secure a formidable lead in the online B2B space.

Maintaining this advantage was a different matter. To keep retention high while growing our user base, we had to focus on not just *quantity* but the *quality* of who was joining. We did this by introducing an authentication and verification system, employing third-party credit agencies to validate the identity and business registration of suppliers. Such investment gave our users greater confidence to connect and negotiate with the prospective business partners they met through Alibaba.

We took this further with Taobao. Unlike professional merchants and business owners, everyday consumers did not have the resources to conduct burdensome wire transfers or bank transactions. This became the premise behind the creation of Alipay—an easily navigable way to process payments for ordinary consumers. This soon became a prominent feature distinguishing Taobao from its competitors; nobody else had provided a tool to address the basic safety and security concerns of online consumers.

Soon after, Taobao added a feature enabling users to leave comments and reviews for the stores on its platforms. This feedback not only gave other buyers more knowledge to make informed purchases but also incentivized better service within our marketplace.

Those tools carried the added benefit of raising the switching cost for merchants who were considering leaving Taobao for other websites. They would be leaving behind their records of successful transactions and reliability. In short, these were their hard-earned online reputations. It is a simple but important reminder of the reality that the surest way to build a loyal client base is by helping those clients succeed.

Mapping Out Your Strategy

An arrow without an archer, a bow, or a target is simply a wooden stick.

In the same way, a company's strategy can never be separated from its mission and vision. Strategy maps are a useful tool for honing the day-to-day operations without losing sight of the bigger picture. As displayed in the chart on page 136, the mapping procedure begins with the customer at the top. Next, we delineate the customer values we have pledged to provide, reiterating each one. Finally, for each value, we determine distinct, specific initiatives that team members can execute. These objectives form a bridge linking the strategy statement to the actual implementation.

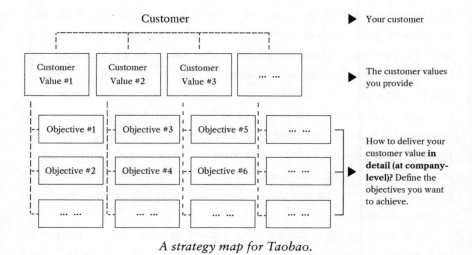

A strategy map for Taobao.

Taobao's strategy statement from 2005 offers an example of this mapping.

Taobao's strategy is to provide a safer, cheaper, more convenient, accessible, entertaining and fun experience to all Chinese consumers by creating the largest online consumer platform and community.

Our task then was to translate each of the four value propositions into concrete features and actionable objectives:

Take "safer," for example. We resolved to enhance server security by instituting greater privacy safeguards to protect against lost or stolen data. We also took on the behemoth that was payment security. One of the roadblocks to more widespread adoption of e-commerce among the general public in China was a lack of trust in sending and receiving payments electronically.

By creating a basic digital escrow system and serving as the designated third party for facilitating customers' purchases—the seed that

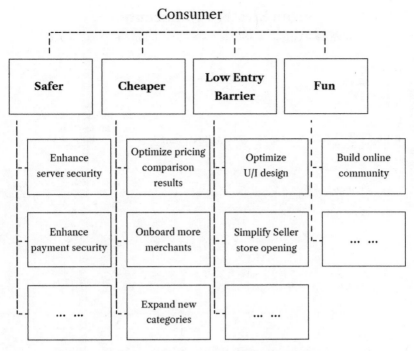

Taobao's value propositions.

eventually grew into the giant Alipay—Taobao both alleviated a bottleneck and honored its first customer value.

Turning Strategies into Plans

For larger organizations, there is the challenge not just of articulating strategies but also of ensuring that these key decisions are communicated and understood at all levels of the company. Here's what that planning and socialization process looks like within Alibaba.

At Alibaba, strategic planning usually begins with informal conversations among the senior executives. This was always Jack's preferred method: to begin by tossing out a bunch of ideas—at times, the more improbable the better—before winnowing them down and refining them through deeper deliberations.

The annual strategy process formally kicks off every August with a group-level meeting, attended by the leaders of every business unit.

From Strategy to Execution

Mission ➤ Vision ➤ Strategy ➤ Execution ➤ Adjustment

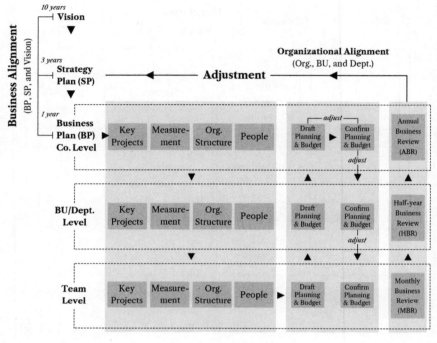

A plan for executing strategy.

Everyone's ideas are presented in an open setting; many of these tend to be opinions about and responses to the developing or projected industry trends. August is a full six months away from the opening of the next fiscal year that these business leaders are discussing, which begins in March. The discussion focuses on a three- to five-year time horizon that could bring Alibaba closer to achieving its ten-year visions.

Zeng likes to say, "Sharpening your axe blade should not slow down the work of cutting the firewood." In other words, strategy planning for the future should take place as a background activity, even as the strategy for present-day business is being executed in the here and now. Zeng points out that, within larger organizations, middle management is a crucial link between the strategic decisions formed at the top and the execution among staff and role players. Middle managers must cultivate their ability to think critically, synthesize ideas, and accurately report the ground-level situations back to executives.

Trial, Refinement, Development

Characteristics of Different Strategic Periods

Trial Stage ▼	Refinement Stage ▼	Development Stage ▼
1999-2001	2002-2003	2004-2006
2007-2008	2009-2010	2011-2014
2015-2017	2018-present	
▽	▽	▽
Let chaos reign	Rein in chaos	Focus
Experiment	Convergence	Discipline
Mission-driven	Strategy-driven	Business model
Bottom-up	Interaction	Top-down

The trial, refinement, and development stages of Alibaba.

Of all the concepts introduced so far, strategy is the most dynamic and iterative. Firms like Alibaba operate within a rapidly changing digital environment, with shifts in technological innovations, consumer tastes, competition, and regulatory policies. Without the ability to quickly understand new developments and adapt, Alibaba could never have survived in this kind of market. Over time, this process of adaptation and evolution settles into a kind of natural cycle, defined by three distinct stages: trial, refinement, and development.

The *trial stage* is defined by discovery and experimentation. During the first couple of years after Alibaba was founded, the budding venture tried out many different ideas. In hindsight, it was lucky that some of these floundered and quickly got buried. For example, China back then had already become the world's largest producer for certain commodity products, and at one point, we had the idea of building exclusive online marketplaces for local governments in highly specialized product areas like socks or neckties.

Another idea that we eventually shelved was creating an online logistics clearing house to coordinate global shipping and export flows. We even looked into trying our hand at organizing offline trade shows. All of these ideas, in their various ways, aligned with our mission to make it easier to do business anywhere. Most of them were suggestions coming from the bottom up, which Jack and the other cofounders welcomed, reflecting the free and occasionally chaotic voyage that ideas could take through a trial-and-error process.

Next comes the *refinement stage* in which the disorder and free-flowing environment is reined in and managers apply some discipline to focus the company's efforts.

Following a number of false starts and thwarted plans, Alibaba made a decision in 2002 to focus on its most viable business model at the time, China Gold Supplier. As one of our few paid products that was getting traction, the Gold Supplier program, a paid service offering a company website with a company authentication and verification check, addressed one of the biggest chokepoints in China at the time—simply getting online with a credible profile. We believed in the offering and devoted our entire sales team to getting the word out. Dispersing around the country, sales representatives went door to door, introducing Alibaba and educating SME owners on the benefits of creating a legitimate online presence that could help their business reach buyers around the world.

Alibaba was teetering toward bankruptcy by this point, and yet we shut down all other business-generating ventures and pooled our resources into this one product. Decisions made during the refinement stage should be driven by your company's strategy. We had determined that China Gold Supplier served a defined need within the market, and we committed to it. As Alibaba strove to reach break-even, the Gold Supplier program provided cover through bleak times and eventually carried us onto solid ground.

Having weathered a harsh storm—not the first, nor the last—we arrived at the *development stage*, the final phase in our progression.

During development, the aim is to streamline your proven business model and fortify it with discipline and specifics across the organization. From 2004 to 2006, we began to standardize all the procedures and workflows related to our Gold Supplier product to ensure replicability and rapidly scale. Our sales representatives had developed a far-reaching reputation for their tireless work ethic and dedication—the legendary *tiejun (铁军)*, or iron soldiers, as we liked to call them—and we codified the steps to their success into our sales training. This system would not only become well established within our Chinese markets but would also serve as the foundation for our international expansion into markets from Vietnam to India all the way to Turkey.

A Virtuous Cycle

Reflecting on that time, I'm convinced that our advance through this cycle is what propelled us to our first IPO listing in Hong Kong in 2007. At that time it was a momentous step, as we started from nothing to create $10 billion in market capitalization within eight years. Reaching such a milestone was a testament to Alibaba's hard-earned maturity through lean years and crises. Yet rather than just trying to hold on to what we had achieved, the company circled right back to the trial stage, throwing itself into the chaos and excitement of new ideas. That entrepreneurial instinct was in our DNA.

Even as Taobao and Alipay were coming into their own, we entered the portal business, experimented with new initiatives like eTao, an e-commerce search engine, and Alisoft, a software-licensing platform, and launched a suite of other productivity tools. This willingness to take on new challenges has been an important factor in Alibaba's ongoing innovation and evolution over the past two decades.

This was the case even after our landmark IPO in New York in 2014. Jack and the leadership celebrated the achievement and then promptly began plotting the next additions to our ecosystem and product offerings.

As Jack and the team researched the markets and sought new ways to fulfill our mission and vision, they concluded that our customers would benefit from services in the health and happiness sectors, which included entertainment and media businesses as well as health care. We began to launch or acquire companies in the film, music, online video, and e-book industries. We also started collaborating with medical and pharmaceutical companies.

All of these initiatives, testing out avenues for business expansion, ultimately formed a new trial stage, which required that we define new strategies. In time we assessed which projects had gained traction and which ones should be shut down—the refinement stage.

High Hands, Low Eyes

Much too often if you talk to senior executives and managers in a company, they will state with confidence that their strategy is crystal clear. They are not intentionally trying to mislead; they spend a lot of their time discussing and planning the strategy and will frequently say the issues reside within execution. Then, when you meet with midlevel staff members a layer or two below, they will often say that their biggest concern is not knowing what the strategy is. They may say they have a good command of their own responsibilities, but they express little understanding of how they fit into the company's wider direction or objectives.

While consulting with different businesses over the years, Zeng has so frequently run into this conflict between what senior executives claim about strategies and midlevel managers understand that he has often skipped talking with executives. To develop a feel for how a business was run, he would go directly to the rank-and-file staff and discuss issues with them. Reflecting on such cases, Zeng adapted a Chinese proverb: "high hands, low eyes." This expression is usually used to describe someone with grand ambitions but meager abilities. Zeng, however, put a positive spin on this old saying, emphasizing the importance both of setting ambitious goals while *simultaneously* ensuring engagement and communication with people at the operating level of companies.

Jack once compared a company's strategy map to the skin of a beast: "You have to find the point where one stab with your knife can bring out the blood." Within any market, true opportunities are fiercely contended. As soon as your competitors smell blood, they will attack. If your company can only engage in hand-to-hand fighting and has no serious weapons, Jack said, you can't break the beast's skin.

Carefully devising a company strategy from a company's *want, may, can* to the strategy statement's *who, what, how* and then mapping that strategy helps unlock the company's path forward. Zeng likes to call this the dynamic upward spiral.

Every new enterprise is typically better suited at locating opportunities than capitalizing on them. The potential is greater than the company's initial capabilities. But as the company develops its operations and seeks to seize those opportunities, the capabilities grow with the experience. If the company survives, those capabilities then build into strengths, and they help steer a company toward the next opportunity, which will further improve capabilities.

Each step gives the upward spiral another push, strengthening the sense of mission. A key management challenge in this process is building an organization that can sustain the upward movement and get staff to cohere around it.

7

Organization and People

Heart, Head, and Hand

What is an organization?

Formally, Alibaba defines an organization as an entity—such as a company, an institution, or an association—comprising one or more people who are coming together for a particular purpose.

The key words in this conception are *people* and *purpose*. No matter their size or complexity, people and purpose are always the raw ingredients that make up the organization. They are the elements that meld together in a collaborative whole to achieve something useful and important. That has been true for many years, of course, but the rise of the digital economy has reshaped how those elements interact and express themselves, how those people serve their bigger purpose, and how they work together to get there.

This realignment has been particularly pronounced in the realm of consumer interactions. As consumption has shifted increasingly from face-to-face encounters to digital shopping, organizations have come to realize the importance of user *experience*—the ease, accessibility, and "stickiness" of the websites and apps consumers use. At the same time, consumers are no longer concerned just with the quality of products; they are also looking to see if a business aligns with their *values*, particularly younger generations. They want to know if companies hold and practice the same values in areas like fair labor and environmental sustainability, for instance.

Because buyers have more ways to express their tastes and preferences, companies must become more responsive. Through social media platforms, review channels, product sponsorships, and ambassadors, the communication between buyer and seller has never been more engaged or interactive. Many firms have developed cocreation models to receive customer input for new product lines or services, further deepening this engagement. This trend requires that organizations become more outward looking, more open to this communication and feedback, and more capable of responding to it.

Also, the internal structure of organizations is changing, with traditional boundaries becoming blurred. Techies have long resisted the rigid hierarchies found in sectors like finance and law. In response, many Silicon Valley organizations have promoted more laid-back environments and flat organizational structures with more fluid operating styles. It is no longer the rule that teams operate narrowly within departments and specialized silos, fixated on just their own tasks. Teams at every level of the organizations are now encouraged to share information, collaborate, and cooperate with one another, fostering an environment that promotes innovation.

While these shifts have had a profound impact on the nature of corporate organizations, some important principles remain unchanged. Both freshly launched start-ups aiming to catch the latest digital trend and older businesses face similar challenges.

First is the question of purpose—*Why do you exist?*—which a company's mission statement must answer.

Second is the matter of direction—*What's the overall plan?*—which vision statements, strategy, and business plans must address.

Third is the matter of execution—*How do you get it done?*—which involves designing the proper organizational structure, recruiting the right people, and motivating them to perform well. How you respond to these questions will be critical to your organization's success.

From Business Maps to Organizational Plans

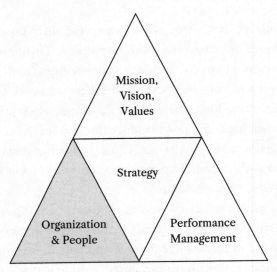

Organization and people form a key part of the Tao of Alibaba.

Organization is the first of two components in every company's capability for execution. In our Tao of Alibaba triangle, we can see that this component, along with performance management, forms the base of the framework, supporting strategy in the middle, which flows out of the mission, vision, and values at the top. In Chapter 4, we referred to organization and people as the *shu* (术) or techniques and tactics for implementing the Tao of Alibaba.

"Business leaders are the first architects of organizational structure," said CEO Daniel Zhang in a talk to a group of young Chinese entrepreneurs a few years ago. His statement was intended to correct the misconception that successful business leaders are solely focused on managing their companies' revenues and expenditures. In fact, Zhang emphasized, a major responsibility for any leader is building the right organization and properly aligning everyone within it.

Designing an organization's structure begins with the strategy plan. The aim is to lay out all the various functional departments, the different personnel roles, and the resource pools available to support and enhance the strategy for achieving goals.

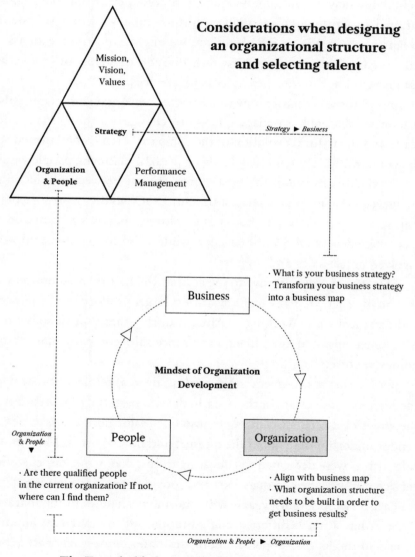

The Tao of Alibaba and organization development.

There must be a coherent logic to an organization's layout. For example, consider a founder who sets his sights on creating China's leading data platform for facilitating smart logistics. If a review of his company's organizational structure reveals that most of his resources are devoted to managing physical warehouses, transport, and procurement, with only a handful of product and engineering specialists in the technology department, then it quickly becomes clear that the organizational structure does *not* align with the company's strategy. Instead, to better achieve his overall objectives, we might expect to see an asset-light organization with a robust tech division to shape an innovative data network for more efficient logistics services.

Importantly, no matter how well designed, an organization must also remain flexible, able to adapt quickly to changes in the market landscape and the natural evolution of the company. I remember around the time of our 2007 IPO in Hong Kong of the B2B platform, people would often ask Alibaba about the breakdown of staff between product and engineering versus sales-related staff. Our COO would surprise them by stating that out of twelve thousand employees, nearly six thousand of them were devoted to front-line sales, while another few thousand were involved in tele-sales and support.

This was counterintuitive to me at first. My natural assumption was that an internet company would feature a larger proportion of engineers and technical staff. Working at Alibaba during that time, it sometimes felt like we operated more like an insurance sales company than a technology venture.

But looking at our operational needs in the marketplace at that time, the logic became clear. In those early days, our staffing favored feet in the street over tech specialists because our major hurdle was a lack of vendor understanding about the internet and how to use it. Most of our sales teams were dedicated to educating the market, while our customer service members fielded questions and provided ongoing training.

Fast-forward twenty years: with enormous changes in the markets, Alibaba has a much different configuration. Of the over two hundred thousand employees, 44 percent of the staff are engineering related compared to 11 percent in sales and customer service. This adheres much more closely to what you would expect from a Silicon Valley company. Though the company's organizational structure and personnel composition have changed significantly over time, they have always reflected the company's goals within its strategy and business plan.

The significance of a sound organizational structure typically emerges under two different scenarios. The first is during the creation of an entirely new company, a stage that all start-up founders must tackle. But mature firms, too, will need to address the challenges all over again when launching a completely new business line, as was the case, for instance, when Alibaba established Alibaba Cloud in 2009 and Cainiao in 2013.

The second case is during restructuring. Sometimes new business needs arise, necessitating a shift in the company's organizational makeup. But rather than scrapping a firm's structure, a more efficient—but no less difficult—approach is to redesign what already exists.

I recall in January 2013 reading Jack's all-staff email announcing one of the company's most drastic restructurings up to that time. Jack revealed that preparations had begun in the summer of 2011 for transforming the company's three main business units into a more agile organization comprised of twenty-five smaller groups.

"This is the most difficult transformation in Ali's 13-year lifespan," Jack wrote.

And yet, he noted: "This is not an occasion where we are making changes because there has been a problem, or because they are overdue. We are doing it because we must face the future we believe in, because we are venturing into territory no company has ever ventured into before."

He then explained the concrete motivations behind the broad restructuring. "We hope to give more young Alibaba leaders opportunities to innovate and develop," he wrote. Jack encouraged everyone to focus not on their "own interests and KPIs" but to strive instead to maximize the overall success of the larger business ecosystem, fostering a more open, transparent, and collaborative environment.

This was an extreme reorganization in many respects, and it required courage and commitment. For many of us, the existing structure felt safe and reliable. After all, it had succeeded in getting us *this* far. Others worried that breaking the company up into so many smaller components was a recipe for chaos. And some felt the timing was wrong: With market competition heating up, should we really be spending so much time on this?

Yet, in time, the decision paid off. In the years that followed, the company grew rapidly, and the restructuring pushed the company to become more resilient and adaptable.

"Of course, no organizational structure is perfect, capable of solving all problems," Jack admitted in his staff email. "Any newly-born organization is like a baby—not yet grown into its appearance and full of problems. But Alibaba's 13 years of experience prove that our people can always devote themselves to making the necessary changes, ensuring that every one of our transformations exceeds our expectations."

This was, in the end, a sign of organizational health, a result of the three-H model we used to sustain wellness.

Assessing an Organization's True Power Through Heart, Head, and Hand

The human body is remarkable. It can withstand shock, absorb damage, recover, and adapt to extreme environments. When we exercise and train our bodies, we undergo a process called muscle hypertrophy, where we expose our muscles to microtears that, when followed by proper recovery, allow our muscles to build back even stronger.

At Alibaba, we have a saying: *A healthy organization can turn impossible to possible.* In the same way that healthy minds and bodies have powered humanity's greatest achievements, from scientific breakthroughs to feats of athleticism, there are few limits to what a healthy organization can do.

Lucy Peng has worn many hats since 1999, but possibly her most influential role was as Alibaba's very first CPO, where she devised a framework for assessing the health of the organization. First, as Lucy suggests, imagine you are a doctor performing a simple evaluation on your organization by answering the following questions on a scale from one to five:

1. Does your organization *understand* and *believe* in your mission, vision, and values?
2. Does your organization have a clear understanding of *what* to do and *how* to do it?
3. Does your organization have the ability to *achieve business results*?

These scores add up to a diagnosis of your company's health and energy levels. In the same way that a high-energy person will complete her tasks and come up with new ideas, a high-energy organization can overcome challenges and achieve objectives. Conversely, a person (or organization) with low energy may suffer setbacks and create new ones for himself and may be more vulnerable to illness.

To gauge a company's energy levels and to be better prepared for improving them, Lucy introduced the three H's.

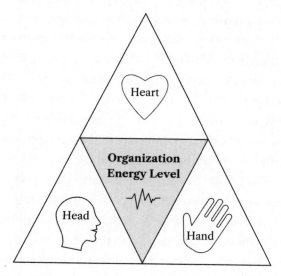

The three H's of a healthy organization.

Heart: The Belief System Powering It All

An organization's heart power is a measure of what its members truly believe, what they are passionate about. Lucy likes to describe it as "the temperature, the air, the intensity you can feel." Throughout his career, Jack has proved many times that a founder's passion for the company's mission is palpable and spreads quickly. It inspires others, stimulating their passions.

Heart power comes from an organization's mission and vision and should be a reflection of its values. Many start-ups are able to spring forward after launch, full of ambition and energy. But as they mature, the execution can become mere routine and difficult and can begin to take a toll on team members.

In some of her training sessions at Hupan University, a prestigious Chinese entrepreneur academy established by Jack, Lucy cited the example of a female entrepreneur in charge of a rapidly growing digital platform. Though she was very driven and capable, the grind of being an entrepreneur trying to maintain growth grew heavier, confronting her with complex difficulties in trying to manage investors, competitors, and her own employees. Inevitably, she began to question herself: *Why am I doing this? Why am I working so hard? What does my company exist for?*

In such moments, your beliefs as an entrepreneur and the leader of your organization must be capable of sustaining you. Being motivated simply to make money may carry you through the early stages, but that rarely lasts long. More holistic beliefs—the wish to share a beneficial product or, as at Alibaba, the aim to help people have a better life through business—often generate better and far longer-lasting results.

Alibaba decided to significantly expand its international offerings a year after its US IPO. We recognized the tremendous potential of such a decision as well as the difficulties. First, we had to identify the right talent to run our international team. We managed to recruit the former vice chairman of Goldman Sachs to spearhead the effort as our president of international business, while many of our country managers came from the executive ranks of multinational internet firms like Google, Amazon, Yahoo!, and Expedia. With that diverse pool of perspectives, it became even more important to ensure trust and common understanding between Alibaba's leadership in China and these leaders in our international operations.

I was tasked to advise this stellar group as we embarked on our globalization journey—no small task given the breadth of everyone's experience and the different ideas each person brought to his or her position. Success hinged on our ability to clearly articulate to our global staff what it meant to be a dedicated Alibaba employee, while at the same time making sure our Hangzhou-based executives in China were open and willing to listen to the perspectives of our European, American, and Asia-based colleagues.

Take our business in Italy, for example. Alibaba faced the challenge of serving Italian companies looking for markets to sell their products, in particular SMEs, while confronting the difficulties of increased competition from Chinese manufacturers going after those same markets. I admired how Rodrigo, our senior executive in charge of Southern Europe, sought to assist these companies by helping them identify areas where Alibaba could support them. He focused on the growing demand in China for Italian luxury goods—Gucci, Armani, Prada, Tod's, Maserati, Alfa Romeo. All of these iconic Italian brands began to embrace e-commerce in new and innovative ways. He also spied a major opportunity for importing Italian foods for Chinese consumers newly eager for tastes from abroad.

At the same time, Chinese tourism was rising sharply across Europe, but spending was limited by rules regulating the amount of cash these tourists were allowed to carry. To address this constraint, Rodrigo and our other international teams worked to expand acceptance of Alipay in popular European destinations to facilitate mobile payments and reduce the reliance on cash transactions by these Chinese visitors, to the benefit of European merchants.

Rodrigo also aimed his efforts inward. He frequently organized internal company events, from documentaries showcasing Alibaba's history and values to team-wide discussions. Rodrigo understood that team members who knew *why* they were important contributors to the Alibaba mission would be more motivated and serve as better ambassadors for the company. He also worked to introduce some of the Italian operation's passion and culture to Alibaba's China offices, once even setting up an authentic Italian pizza concession stand in the cafeteria of our Hangzhou headquarters so staff could appreciate the value of Italy's contributions to the company.

Lucy often said that heart power is the first and most important of the three H's. "Heart power should inspire and excite others," she said.

"Others should be able to feel the warmth and value of the things you are doing."

Rodrigo was one excellent model of heart power in action. During 2020, when the coronavirus pandemic was locking people in and infection rates were soaring, particularly in Italy, the Italian team recorded a virtual musical performance of employees from all over the company, singing an Alibaba Italy song. He also created a new video show format called " 30 Minutes with Alibaba Italy" for the employees to meet special guests to talk about life topics and provide inspiration and hope amid challenging times. Though small steps, these actions raised spirits for many of us, confined to our home offices and in need of some warmth and amusement.

Head: The Brains of the Operation

Heart power is the passion and spirit that inspires team members and partners to march forward and make it through even the most difficult of times. But such energy needs guidance, and that direction comes from the head. Head power is the rational thinking, logic, theory, methodologies, frameworks, critical analysis, and professionalism needed to identify the real customer value and to deliver it in a way that helps customers solve their problems and advance the company's mission.

An organization's head power in the modern digital era is more important than ever before because of unrelenting competitive forces. In some respects, it has never been easier to launch a start-up, with abundant funding opportunities, but navigating a path toward profitability has never been more arduous because of the intense competition. With so many ambitious entrepreneurs willing to do anything to gain even a slight edge, the markets have become minefields.

Though Alibaba had to overcome many difficult periods, it also had the advantage of championing what were then novel solutions to real consumer needs. We were swimming in a "blue ocean," so to speak, where the challenge was explaining what we did and persuading potential customers that it was valuable, not fending off competitors. Today's booming digital economy is a "red ocean," filled with powerful competitors and innovators. In this environment, the ability to assess your company's strengths and plot the proper strategic choices can make the difference between surviving or bowing out early.

With DingTalk, Alibaba's workplace messaging app, cofounder Wu Zhao was able to apply logic and strategic thinking to turn the company's previously unsuccessful consumer instant messaging service, Laiwang, into a tool that revolutionized the business productivity space in China. Wu observed that many Chinese companies, regardless of their size, lacked a cheap, effective, and easily accessible solution for team-based communication. Large corporations usually had their own intranet system for sharing files and documents, but the servers were frequently difficult to navigate on mobile devices and required cumbersome network log-ins to perform even basic tasks.

Small and medium-sized businesses rarely had the budget to establish a proprietary corporate server, while consumer messaging apps like WeChat lacked the file management capabilities businesses need. The solution Wu and his team came up with was to leverage Alibaba's knowledge from years of experience as a business marketplace, a deep understanding of the usual pain points around communication and productivity, to build an all-in-one app to manage all those needs efficiently. It became a top choice for businesses and is now the most widely used enterprise messaging software in China.

During the COVID-19 pandemic, DingTalk's enhanced functions made it far easier for many businesses to transition to remote work, as well as schools, earning it a recommendation from UNESCO as one of its preferred distance-learning apps. It also became the tool of choice for medical organizations from more than forty-five countries for sharing COVID prevention and treatment information, involving eleven languages. Wu and his team's analysis and reasoning made DingTalk not only a useful product but an *essential* one. That was Alibaba's "head" at work.

Hand: The Power to Make Things Happen

Hand power is the final, crucial piece of an organization's health. This is your company's capacity for execution. This requires logistical planning, accuracy, flexibility, and sometimes physical endurance.

In 2014 we launched a program to help entrepreneurs in the Chinese countryside set up e-shops to sell goods to consumers, calling it Rural Taobao. This was an important but challenging project that required visits to many out-of-the-way villages. We said that the team members needed the drive and spirit of the *tiejun* or iron soldiers sales force who had worked extremely hard to get the business off the ground in Alibaba's fledgling B2B days.

As we did in our core business units, the Rural Taobao managers set clear KPI targets at the national level and then customized and applied them at the provincial level, all the way down to counties and villages. The team members were motivated by a sense that the project was serving a valuable mission of assisting villages that had often been left behind in China's dash to modernization, and they executed the strategy energetically.

On the ground in many of these remote outposts, Alibaba benefited by partnering with small businesses and young workers who, in many instances, had left home and then returned to try and find success in those familiar surroundings. We helped them set up local stations that served as logistical and educational hubs for villagers who wanted to participate in e-commerce.

The speed and scale of our implementation was remarkable, reflecting both the appetite for such digital market access among the local populations and the execution of our team members. Rural Taobao locations rose from 212 villages at our initial 2014 launch to more than 30,000 by 2018. During my visits with our Rural Taobao teams from Anhui province to Inner Mongolia, I was struck by the sense of duty and discipline—the *hand power*—evident in everyone's efforts. That ethos resembled the discipline and commitment of a real army, marching forward to complete the mission.

Bringing the Three H's Together

One of the benefits of the three-H model is it allows every organization to assess its relative strengths and weaknesses within the framework. Just as a doctor takes into account various sets of symptoms in making

Type of Organization	Organizations	Heart (Purpose)	Head (Know-How)	Hand Execution
Type A	**Star**	✓	✓	✓
Type B	**The Lost**	X	✓	✓
Type C	**Startup**, not enough business experience	✓	X	✓
Type D	**Strategist**, lack of execution, couldn't get things done	✓	✓	X
Type E	**Daydreamer**	✓	X	X
Type F	**Planner**	X	✓	X
Type G	**Executor**	X	X	✓

Assessing a company's health is a lot like a doctor assessing a patient.

a proper diagnosis, the leaders of an organization can look to the chart above to get a holistic view of the company they are managing—its tendencies and potential weaknesses.

Take Type C, for example, an organization that exhibits plenty of heart and hand power but is deficient in head power. This is the typical mix observed in most start-ups, as they burst out of the gate full of passion and a strong work ethic but little of the know-how that comes with experience.

Meanwhile, a Type B organization features head and hand power but is missing heart power. We refer to companies in this category as "lost" because without a compelling purpose the tactics and execution may be useless. Many second- or third-generation family businesses might fall into this bucket; the founders of the company established it with a strong sense of purpose, but success and changing conditions over time may have diluted it.

Some organizations may be brimming with mission and drive, but they don't know how to apply this energy and get things done. These are the *daydreamers*, a category that includes many student entrepreneurs at the beginning of their journeys.

Other organizations can fall into the trap of becoming obsessive *planners*, full of time lines and goals but missing the other two vital ingredients to get the business off the ground. Occasionally, companies may excel at execution—essentially following orders and delivering results. But without a head and heart, they risk never expanding beyond a narrow lane.

The goal for every business is to identify and enhance the areas where they are weak—then bring heart, head, and hand powers to the levels required for sustained success. For this task, we can consult the three circles of developing an organization.

Developing Your Organization Through the Three Circles

After applying the three-H model to diagnose the health and energy levels of an organization, the next step is to devise a proper diet and fitness plan to improve all aspects of performance. For this, we implement the three circles framework.

At Alibaba, we define a *healthy* organization as one that has the will, the ability, and the capabilities to deliver on its business objectives. The *will* is a measure of the organization's heart power, the *ability* comes from the head power, and the *capabilities* are mechanisms that relate to hand power. You can think of the three circles as a special set of exercise regimens, specifically tailored to optimize the health of the operations.

The Inner Circle: Culture

The innermost circle represents your organizational culture, reflected in the values, behaviors, and expectations of your employees. Your mission, vision, and values form the core of your organization's culture, but

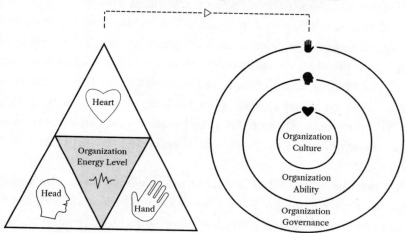

Company Health Check ▶ Actions Required

The three-H model can allow us to devise a plan to
upgrade company performance using the three circles framework

*The three H's and the three circles come together
to form a healthy organization.*

culture extends beyond MVV. Culture is a complex, evolving thought process, shaped by many inputs and influences. It can be understood as a sense of belonging, the appropriate patterns of behavior, wide and deep enough to encompass each individual employee at the company.

During the initial start-up phase of an organization's life cycle, the founding individual or team typically make an outsized imprint on their company's culture. When Jack founded Alibaba in 1999 with his seventeen cofounders, the group dynamic was distinctly collegial: Jack carried his interactive teacher persona into all his meetings while his cofounders were fellow instructors and colleagues. New hires, once socialized into the firm, were referred to as *tongxue* (同学), meaning "classmate."

Fast-forward to today, with headcount surpassing more than two hundred thousand. Employees still address one another as classmates— including the security guards at the campus gates and the cafeteria workers. These small gestures help instill a sense of community within the company.

And with his love for martial arts novels, Jack has infused some of that *wuxia* sensibility into Alibaba's culture. Jack often invoked anecdotes from these popular stories and folktales to inspire a heroic spirit among the staff, shaping a company ethos of helping the underdogs (the SMEs) just like the protagonists in his favorite stories.

The practice of having employees select nicknames from the *wuxia* stories reinforced this culture. Jack adopted the nickname Feng Qingyang, after the elder swordsman of the Mount Hua Sect in Jin Yong's novel *The Smiling Proud Wanderer*. Wise and clever Feng is the reclusive master who agrees to teach the novel's protagonist, Chong Linghu, the Nine Swords of Dugu fighting technique. With this vital training, Chong becomes one of the most famous and respected swordsmen in Jin Yong's anthology.

Daniel Zhang, our CEO, selected the name Xiao Yaozi, the brilliant kung fu master from Jin Yong's novel *Demi-Gods and Semi-Devils*. Xiao creates the "eternally youthful style of fighting" and establishes his own kung fu sect, which attracts some of the greatest fighters from around the world to stand up for noble causes.

These lofty tales and legends may sound childish, but they provided rich imaginative reinforcement for the culture among Alibaba's staff and team members. The causes were just, the sentiments pure. Though early staff members long ago grabbed the nicknames of the more well-known characters in Jin Yong's stories, team members and classmates can still adopt inspiring nicknames to bolster their motivation and sense of moral purpose.

Another unconventional custom at Alibaba is the requirement that all new employees learn how to do a handstand during their orientation training. This ritual harks back to the cold winters in Jack's Hangzhou apartment when Taobao was first launched. In those days, the employees usually stared into and tapped away on their computers from morning to night. To prod the team members into taking more breaks and clearing their heads a little, one of the engineers came up with the idea of doing group handstands against a wall.

When Taobao was launched in 2003, the practice took on new meaning: doing a handstand, we said, meant looking at the world from a different perspective. At the time, as Taobao was competing with eBay in the consumer market, this practice became a critical reminder of the importance of innovation, that we needed to approach our business and strategy from new, unexpected angles.

Another custom grew out of Alibaba's vision statement that the company strives not simply to be a big and influential business, but also a *good* one. That is why, beyond doing good through business, Alibaba seeks to make positive social contributions through many programs and policies. Every employee, for instance, is encouraged to perform at least three hours of community service each year. The point is not to mandate minimums but to remind everyone of the rewards of serving others. In reality, Alibaba employees frequently end up volunteering far more than the minimum three hours and benefit from forming a link between their office duties and a commitment to supporting their communities.

Alibaba also commits to donating 0.3 percent of its total revenue each year to the Alibaba Foundation, a philanthropic organization operated by the employees themselves. Over the last six years, the foundation has donated money to support issues like rural education, the environment, and wildlife protection, as well as women's empowerment. It frequently partners with international organizations like the UN, the World Bank, and others to amplify the impact of its grants. These initiatives inform Alibaba's culture and provide a supportive ethos in line with the company's vision "to be a good company that will last for 102 years."

The Middle Circle: Ability

The middle circle represents organizational ability, or the mix of skills and competencies that get everyone pulling in the same direction, completing necessary tasks in the right order and propelling a company to succeed. While an organizational culture creates a shared sense of purpose and belonging, organizational ability describes the key characteristics that will enable a company to deliver on that purpose and achieve results. These traits are closely linked with the type of organization you run: they are defined by quality rather than quantity and will rarely number more than one or two.

Different kinds of organizations will prioritize different organizational abilities. The commander of a military unit, for example, might value unit discipline, with everyone carefully aligned within a highly efficient system of centralized command. A kindergarten teacher might emphasize kindness and patience. Government leaders may recognize the need for transparency and equal justice. Filmmakers may prize creativity and focus. In all these cases, the most important organizational abilities change to support the purpose.

As a company progresses through the life cycle from start-up to incumbent, the organizational abilities also progress to suit the stage of development. Young companies may focus on generating passion or initiative, only to find as they mature that the capabilities they most need are trust and education. Alibaba's major business units have honed and prioritized different organizational capabilities at various stages along the journey.

When Taobao first launched in 2003, its managers prioritized innovation in the effort to battle with eBay. First, the platform needed a strategy to bring more Chinese consumers online, then Taobao had to figure out the actual products and services it wanted and needed. Eventually, the managers had to deepen consumer trust to turn them into loyal users. At each stage, the managers needed to understand their users as deeply as possible and adjust the organization to deliver on consumer needs.

In those early days, all Taobao employees submitted a weekly report reviewing their performance and assignments. The management then decided to add an additional question, asking employees to list three areas of the larger business they thought could be improved. Team leaders were tasked with assessing this feedback and choosing the best three suggestions, which they submitted to their own superiors right up to the CEO. In this way, Taobao's leaders could access the perspectives of their front-line staff, leveraging this knowledge to inform adjustments and new innovations.

Sometimes, Alibaba has sought to redirect an organizational ability through indirect, or symbolic, means. In 2009, during the early days of Alipay, Jack had suggested that Simon Hu, a senior Alipay executive, start a new microloan business to serve SMEs. The decision was made to utilize Alibaba Cloud's infrastructure to power this new start-up, but in the process everyone realized that the system was plagued by constant outages and bugs. Cloud computing technology was still emerging, and though the innovations in the field were exciting, the prospects for achieving profitability were far from certain.

The team decided to persevere in using Alibaba Cloud because only if they had a successful use-case could they then prove that the cloud infrastructure was viable for broader use. Hence, sustaining the endeavor necessitated another kind of organizational ability—persistence. Simon decided that, to promote this faculty, they would add something unusual to the agenda for the company's annual leadership retreat. They called it a "Journey to the West"—it was a grueling 122-kilometer hike through the sandy plains of the Gobi Desert, taken over several days.

The leaders had four operating principles: to believe, to act, to persist, and to exceed. On such a trek there was no way to turn back because any individual left alone would likely perish in the harsh conditions. The only direction was forward, together.

As a result, persistence became one of the most critical aspects of the team culture. Following this experience, the Alipay leadership team believed anyone who conquered this long physical ordeal would understand more deeply the kind of persistence needed to meet the Alipay business objectives. Indeed, this cultural trait not only fortified the team's resolve but also was utilized by Simon to build his new team's spirit when he became the leader of the Alibaba Cloud business. By placing such emphasis on persistence in the culture, it helped drive Alibaba Cloud's successful growth.

The Outer Circle: Governance

Finally, the outermost circle is organizational governance. This is a system that helps determine how an organization makes and implements decisions in pursuit of its objectives. It can include the processes, mechanisms, standards, and rules an organization follows in decision-making. It guides everything from operations to ethics, from compliance to risk, and more.

Jack often spoke of his dismay in seeing once respected, successful organizations such as Yahoo! or Hewlett-Packard decline. The professional managers who replace the founders can harm the companies by prioritizing quarterly earnings and stock awards. Their behavior is not always geared toward long-term growth but rather guided by short-term goals like pleasing shareholders and compensating executives. In other words, when founders retire, their visions often retire with them. A company's direction and purpose becomes diluted, or disappears altogether.

Jack and the other founders feared the same might happen to Alibaba if they did not design and put in place a system to preserve the company's special culture for the long run. So in 2010, they created the Alibaba Partnership. Under this structure, new partners are elected each year. The existing partners nominate new candidates based on their experience, leadership, and commitment to furthering the Alibaba culture. A prospective partner must get the approval of 75 percent of the existing partners to earn a place in the partnership—no small feat. These are regarded as exemplary leaders who are particularly strong in their understanding of the company culture.

In its official statement, the Alibaba Partnership guidelines note that new partners must demonstrate the following:

- a high standard of personal character and integrity;
- continued service with Alibaba Group, our affiliates, and/or certain companies with which we have a significant relationship, such as Ant Group, for not less than five years;
- a track record of contribution to the business of Alibaba Group; and
- being a "culture carrier" who shows a consistent commitment to, and traits and actions consistent with, our mission, vision, and values.

The partnership currently has thirty-eight members, all from different areas of the Alibaba ecosystem.

One aim of the partnership structure is to reduce the common problem known as key person risk, or the risk that a company will be damaged if a founder or other key employee leaves. The partners, it is hoped, offer some protection by building a cadre of talented employees who, among other things, embrace and exemplify the culture. They can also ensure that the company remains focused on long-term objectives, not short-term metrics. The partners have, for instance, the exclusive right to nominate a simple majority of Alibaba's board of directors, subject to shareholders' approval during the annual general meeting.

Some have argued that the structure puts a disproportionate amount of control in the partners' hands, but the aim is to put in place a governance structure that continues to place a priority on long-term objectives and long-term shareholder value.

This aim came up in 2013, in a public letter from Joseph Tsai. He emphasized the importance of strong governance to maintaining Alibaba culture. "For the past 14 years, Alibaba operated with the ethos of helping the 'small guy' to succeed, as embodied in our mission: 'to make it easy to do business anywhere,'" he wrote. "The final governance structure we have selected is to replace founders with partners."

The reason for this, he continued, was simple: "A group of partners who cherish the same culture and ideals is more likely to carry forward our principles and make good decisions for all stakeholders with a long-term view. And in the decade to come, those partners will be guided by these principles when grappling with inevitable disruption and competition."

Aligning Culture, Ability, and Governance

To make the three circles—culture, ability, and governance—work for you, there must be alignment within your organization: what you say (about your company culture) must be reflected in what you do (ability and governance), which must then be communicated both internally and externally, in effect creating a virtuous cycle that can capably reinforce itself.

Alignment — Culture, Ability, and Governance

Behaviors	Internal Communication	External Communication	Communication
	Record, edit, organize, research, and publish (articles, video, livestreaming)		
	Activities		What you do
Organization Governance	Mechanism Partner and Sr Management mechanism, Performance mechanism (361), Recognition mechanism (C&B plan, Red and Rotten Strawberry Award, etc.), HR mechanism (hire and fire, code of conduct)		
Organization Ability	Organization Ability		
Organization Culture	Culture		What you say
	Mission, Vision, and Values		

Culture, ability, and governance must be aligned in a healthy organization.

People Strategy

People Strategy
should align with business strategy and organization

· What is your business strategy?
· Transform your business strategy into a business map

Business Strategy

Mindset of Organization Development

People

Organization

· Align with business map
· What organization structure needs to be built in order to get business results?

· Are there qualified people in the current organization? If not, where can I find them?

People strategy must align with business strategy and organization.

With proper alignment among the three organizational tenets of your company, we come to the final component of management strategy—your organization's people strategy.

Your company's business strategy, discussed earlier, determines the most effective organizational structure. You must then decide what sort of talent and people are required to fill those positions and deliver the desired performance.

Here there are two main considerations to weigh. The first is to take the measure of the size and the quality of the local talent pool. Managers must ask themselves, can they attract and hire the right kind of talent, employees who can capably fulfill their roles and help implement the company's larger strategy and business plan? If the answer is yes, then the manager's first step is to clearly define the skills required and begin identifying suitable candidates.

However, frequently—especially for start-ups—the answer may be no. The right talent does not exist in the local market, or the manager doesn't believe he or she will be able to attract talent, either because the company is too small or because it does not have enough resources. In this case, we arrive at the second scenario, in which the manager must assume the responsibility of training staff internally to meet the organization's needs. While the growing popularity in remote work may alleviate the availability challenge somewhat, for new ventures or those companies unable to afford top talent, a certain level of internal training is still required. This is typically a more committed process, requiring both more time and greater effort, but it is not uncommon among small firms or those operating in frontier markets where the local conditions may still be underdeveloped.

Organization: The Key to Resiliency

Throughout my first tenure at Alibaba during the company's early years, I recall feeling frustrated with the constant changes—big and small—that disrupted work patterns. Our business strategy might be suddenly diverted in a new direction midstream. Product launches could be delayed or even canceled with hardly any warning. And, most unnerving of all, major changes in the company's structure were announced sometimes multiple times in a single year.

It was jarring. Just when I had settled into a new desk and started getting to know my teammates, I would be asked to switch departments and begin my orientation all over again. Most days, I couldn't help but wonder how "normal" companies operated.

Predictably, I enjoyed a more consistent, stable work environment after I graduated from business school and joined the ranks of McGraw Hill, a large multinational in New York. But what I wouldn't come to fully understand until later, after returning to Alibaba, was that such shake-ups and pivots were essential for our need to adapt and thrive in the digital markets. Sure shake-ups were not without costs, often resulting in the loss of those staff not comfortable in a volatile work setting, but since digital markets grow and change at astounding speeds, so must the companies seeking to succeed.

Crucial to achieving this kind of responsiveness and resiliency is the organizational structure. The constant reconfigurations I experienced at Alibaba were necessary for our exponential growth.

At the same time, organizational principles can only get you so far. They are a tool, a support framework. The right structure for your company will help to align employees and amplify their best qualities, but it is the talent of the employees that ultimately determines whether a company triumphs or stumbles. Knowing how to select the right people, then to properly train, evaluate, and incentivize them, is what ensures that a company can execute and fulfill its mission.

8

Performance Management

Going Beyond KPIs

A pril at Alibaba is always a hectic time. Just as teams wind down the fiscal year, management begins implementing plans for the new one. It can be hard to focus fully on either the year ahead or the year behind. And then, on top of all this, April is the month for performance reviews.

I still remember my 2006 review vividly. I'd had a challenging few months devising creative ways to convince international buyers to come onto our B2B platform—a task for which I'd been given barely any budget. Not to mention, few people in the company at the time had international marketing experience. Deep down, I knew that we had worked harder than ever that year. But, I wondered, would my effort on this tough task be rewarded?

The morning of my review, I entered the Motianya conference room at our headquarters feeling nervous. It was one of our many meeting rooms named after the mythical locations from Jin Yong's martial arts novels. These locations were, typically, lush mountain forests or high desert cliffs where the kung fu heroes lived in seclusion, reflecting on life or imparting their wisdom to disciples.

The conference room's name, conjuring confrontations with fate, fit nicely with the purpose of a performance review. I sat at the conference table with my immediate supervisor, Liqi, and my HR manager. I braced myself for the conversation. You could never quite be sure which direction the review would go. More often than not, supervisors would spend less time on all the great things you had achieved and more on what you needed to improve upon and why.

I'd been preparing for two weeks, both by rehearsing my spiel with my colleagues and on my own. I needed to give a summary of my work over the past year, and I had the information down cold. Once I got started, I was presenting exactly as I'd rehearsed, but just as I finished my third slide, a sly smile crept over my supervisor's face.

"Brian, that's fine," he interjected. "We know about the good stuff already. Let's talk about the areas you think you *didn't* do well in this year."

This interruption was not in my presentation plan. Flustered, I tried to regain my composure as quickly as possible and skipped ahead to the end of the presentation, where a few slides offered reflections on areas I felt could improve on moving forward.

"Thank you, Brian, that's very comprehensive," Liqi said. "You're clearly enthusiastic about the work you've done, and you work hard. You effectively motivate your team, and you often share your excitement about Alibaba's mission with your teammates."

OK, so far so good, I thought, feeling a bit of relief.

"The only issue," he continued, "is that you haven't focused on the true needs of the customer. You spend too much time looking outward, trying to build partnerships."

Huh? I thought, wondering where this was headed. *I thought the partnerships were something I'd be congratulated for.*

"But the trade show partnerships you've set up haven't translated into online traffic. Nor have they brought any quantifiable value to our China Gold Suppliers," he said, referring to the Chinese customers who were paying for this service and counted on us to attract more international buyers of their products. "You know at Alibaba we applaud you for the process but pay you for the results."

Seriously? I thought. With the budgetary constraints I was under, as well as the limitations of our technology, I was sure I had exhausted our options in bringing in those international buyers. We'd signed more than twenty partnerships with some of the world's largest international trade shows—no small feat at a time when people barely knew what Alibaba even was! I was sure that, in time, those partnerships would pay off.

The critical feedback took me by surprise. I assumed I had done well, given the difficulty of the challenge I'd been given. I also felt that the company had placed an outsized burden on my team's shoulders and hadn't given us the same level of support as colleagues who were operating in the domestic market. The China-facing marketing teams had legions of people, all working to build their user base. I had three or four staff members for each of my *three* vast regions: the United States, Europe, and Asia. I was given a skeletal team but was still expected to cover the entire world.

But Liqi wasn't done. Given our growth in new China Gold Suppliers, he explained, we needed to be finding new and innovative ways to grow that base of buyers to expand sales for the Chinese producers. Buyer growth could be measured by the number of buyer inquiries being generated through the platform. But if this number wasn't growing fast enough, then our China Gold Suppliers, mindful of the costs of remaining on our platform, were likely to look elsewhere for opportunities.

As frustrating as it was to hear all this, something clicked. It dawned on me that Liqi was right: I had been so enamored by the success of my team in lining up the trade show partnerships, both in the United States at eBay Live and at Hong Kong trade shows, that I neglected to nail down the actual commitments of buyers. I had been fighting so hard for the trade show eyeballs that I had not done enough to bring them to our online platforms.

While it would have been nice to have more recognition for my team's efforts, I came to realize that, in one way or another, everyone was struggling to reach large goals. The teams with sizable staffs had goals of an entirely different magnitude. The level of dedication asked of any employee at Alibaba was high—higher than anywhere I'd ever worked before.

Many of my China-based colleagues were spending weeks on the road trying both to sign up new Gold Supplier accounts and maintain relationships with existing ones. Although they were in sales positions, they took on training and account management and served as general 24-7 management consultants for their clients' e-commerce support, often fielding phone calls and making visits at all hours of the day.

Some regional managers had moved to different provinces for up to a year to open new regional markets, while their families remained back home. Mentally, it was a reality check. In order to succeed here, I needed to put my ego aside and be realistic about shortcomings. It was time to step back, assess, and recalibrate for the next round. That was the message I took from the performance review.

Performance management is the final component of the Tao of Alibaba.

Performance culture is another element of *shu* (术), or skill or technique, required for implementation of the Tao of Alibaba, and it is one of the core components that supports the execution of a company's strategy.

At Alibaba, performance reviews are a cross between a personal therapy session and report card day. They are a time to offer praise, tough love, and compassion—not only to help employees improve their performance but also to help them gain self-awareness around their own strengths and weaknesses.

CEO Daniel Zhang breaks down Alibaba performance evaluations into two parts: results and process. Results can be assessed using KPIs, but managers should use company values to evaluate processes, *how* team members are getting things done. If evaluators disregard process, or only care for process but not results, the worth of a team's achievements is not being weighed properly. Only in concert can these two measures form a meaningful, impactful performance evaluation.

I encountered this myself that year, when my supervisor made sure to praise my hard work, or process, while reminding me of how I'd fallen short of my results. Of course, it was hard to hear the disappointment, but once I put my emotions aside, we focused on ways I could address the problem and how my supervisor and team could support me in overcoming my weaknesses. In fact, as soon as I realized that the purpose of this performance review method was to help me improve, I appreciated the real benefits of the exercise.

As I progressed on this learning curve, I also came to place greater value on my own team reviews. In fact, they became one of the most satisfying aspects of my role as a manager, giving me an opportunity to hear what mattered most deeply to them while assisting their professional development in a meaningful way.

In short, there are two key, interrelated principles that inform performance management:

- Principle one: results are important, but so are the processes.
- Principle two: applaud the process, but don't lose sight of the performance.

Though I had to learn principle two the hard way in that conference room in 2006, I was able to embrace the ways good reviews empowered us as individuals and teams. Managers were obligated to support employee development, but the employees had to take responsibility for their own growth.

To facilitate this approach as an ongoing, constructive process, supervisors must create regular opportunities to communicate with individual employees, coach them, and pass on relevant aspects of Alibaba culture. Employees need to take advantage of these opportunities to understand and implement supervisor feedback and to work with their supervisors to plan their development. The overall organization plays a role by establishing a unified, transparent system for the evaluations and by guaranteeing regular assessments.

Performance Evaluation and 996 Culture

When companies grow, and especially when they grow quickly, they can become more impersonal, losing some of the spirit and intimacy from their early days. Employees can feel less connected to their colleagues and team leaders. They might feel their work has little impact on the company's mission. Thoughtful, high-quality performance evaluations are a way to address this and make sure each individual's needs are being met. They're also a way to make sure employees are being recognized and compensated accordingly for hard work and results.

In recent years, China's booming tech companies have come under fire for how hard they drive employees—what is called the 996 work culture. 996 is shorthand for when a company requires employees to work from 9 a.m. to 9 p.m., six days per week. Alibaba first used the grueling 996 work schedule when it made its push in 2013 to go all-in on mobile.

The competitive pressures were enormous at that time. Tencent's messaging platform, WeChat, was experiencing massive growth, going from little known to ubiquitous in the span of a few years. Up to that time, Alibaba had been focusing on cloud development, but the WeChat success was a wake-up call that we were at risk of missing an even more urgent priority. Alibaba needed to pivot to mobile if it wanted to catch up.

We realized that if we did not push immediately into mobile, we might never catch up: 996 schedules became a necessity and even a rallying cry. Jack spoke in favor of it, noting that 996 dedication had led to quick innovations that might have taken far longer to develop otherwise.

Eventually, we turned the Taobao mobile application into a great success, but this working style certainly isn't for everyone. Jack believes, though, that nothing comes easy in life and that Alibaba was actually giving its team members the opportunity and motivation to create something meaningful.

The 996 approach has morphed into something different since then, however. Other tech companies embraced it, but largely as a way to squeeze more effort from employees. It has become divorced, in other words, from its original intention—a *temporary* system for achieving a specific, critical goal. Now, it can feel exploitative, like demanding that employees run a marathon at a sprint pace. As a result, many regard 996 culture as an abusive management technique, not a path to making something meaningful.

A strong culture of thoughtful performance evaluation can do a lot to combat this. After all, one of the core functions of a performance review is to ensure team members are compensated appropriately for their efforts, whether working at a 996 pace or not. It's also a time for managers to connect with employees on a personal level, to hear about their dreams, their goals, and their frustrations. If 996 is not productive, or not what an employee expected or wants from the job, then that's something important for both sides to discuss and take into consideration.

The Performance Management Circle

At Alibaba, performance management is looked at in two ways: the company perspective and the employee perspective. From the company's point of view, Alibaba is a mission-driven internet company that pursues excellence. Accordingly, its performance management adheres to a concept of "high performance and high return," which is intended to ensure a maximum success rate in meeting company goals.

But Alibaba also pays attention to employee growth and development, helping employees focus on their goals and improve their working methods and using performance appraisals to encourage innovation. In turn, by exemplifying the company's culture and values, the employees are contributing to Alibaba's healthy, long-term development.

All of this plays out in the form of the performance management circle—the annual cycle of setting goals, coaching, evaluating, rewarding, and reviewing goals. As shown in the chart below, the circular nature means that performance-management-related tasks are continuous, forming a regular cycle for employee evaluation. Reviewing isn't just an annual task but an ongoing journey of reflection, designed to cultivate awareness around oneself and one's team.

Performance Management Loop

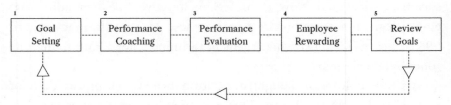

Key components of the performance management process.

Still, it can be hard to think about these performance management concepts in a purely theoretical way. That's because they become more than ideas when they're ultimately so personal. Performance reviews can be charged with emotion. I've seen people walk out of the conference rooms sobbing in disappointment and frustration. Just as many times, though, I've seen employees leave reviews full of excitement and joy. In each case, emotions can get in the way of our ability to think critically about our performance. And that's why the following principles are important. They help anchor us and manage expectations when our feelings threaten to get in the way.

Setting Goals

On the performance management circle, setting goals comes first. But selecting the right goals for measuring performance can be tricky. It should be about more than selecting KPIs. Like everything at Alibaba, goals should be one part of a holistic perspective that includes mission, vision, and values. We like to say that mission and vision provide direction, values are the traffic lights, and KPIs are the dashboard.

Goals translate your overall strategy statement into tangible, three-year objectives. The most effective way to do this is break down the customer values defined within your statement into specific components of your annual business plans.

Previously, I shared Taobao's early strategy statement:

Taobao's strategy is to provide a safer, cheaper, more convenient, accessible, entertaining and fun experience to all Chinese consumers by creating the largest online consumer platform and community.

Here, we can identify making the shopping experience safer, cheaper, and more convenient as key value propositions. The next step is to construct a plan for implementing these objectives in a three-year time frame.

Let's start with making the shopping experience safer. We could create specific projects to achieve those goals, like the following:

1. Enhance the user feedback system.
2. Enhance server security.
3. Enhance payment security.

We would then break these projects down further, into yearly goals. For "enhancing user feedback system," the team could set more granular goals for the first year:

1a. Revamp the system and interface for buyer feedback.
1b. Create a new "friend circle" function, allowing users to see what their friends are buying.
1c. Provide verified merchants' business license information.

Each of these goals would then be assigned even more detailed objectives within the business units, departments, and working teams handling the tasks.

As you can see, there are two major components to all this. At a higher level, we convert our strategy into a business plan. And then we filter down to lower levels by translating our business plans into actionable goals for the teams. Each of those layers requires mapping the business model, key projects, measurements, organizational resources, and then the budget. Ultimately, the resulting goals are used as a foundation for evaluating progress, from the company overall and then down to the individual.

The Overall Objective: Evaluating Performance

Alibaba's mantra of "customer first, employee second, shareholder third" guides everything the organization does—including performance evaluation. Customer satisfaction is the main criteria for determining whether a team or employee's work deserves recognition. Customers refer to both external customers the company serves but also internal stakeholders and colleagues who one must collaborate and cooperate with.

Gathering customer feedback has been baked into Alibaba's development since its start. September 10, Alibaba's birthday, is also designated as Customer Day, when staff and often senior management get on the customer hotlines and listen to the feedback to grasp firsthand what's happening on the front lines. Senior management have been known to staff the customer hotlines as a way of gauging the overall performance of the company—as well as determining which teams are having an impact.

Some employees like the ironic joke that the best way to determine if you are putting the customer first is to see whether Jack is satisfied with your team's performance. Jack does often demonstrate an uncanny ability to sense market sentiments and popular perceptions of his business. But, of course, Jack is one observer, and not everyone can get his personal assessments, so most of us relied on criteria that assessed customer first metrics along two dimensions: ability versus willingness and internal versus external.

There are different ways to measure ability. For sales teams, it's straightforward: sales data are a hard indicator. But for nonsales teams, ability is assessed differently. Let's consider teams responsible for web products and design, for example: they should be evaluated based on the ease with which buyers can search for products or sellers can operate their platform. Though the indicators vary based on the team, they all target the customer experience.

Meanwhile, willingness (or intent) relates to putting in extra effort to meet customer needs. Alibaba encourages employees to take initiative to solve problems, even when they don't necessarily fall under their role's scope or fit in with their KPIs. For example, a colleague of mine on the training team encountered a customer complaint about a certain product. Though it didn't relate to the training exercise we were doing, the employee took the time to find the responsible team and pass on the information.

Internal versus external acknowledges that Alibaba's complex ecosystem consists of numerous stakeholders inside as well as outside the company, all of them "customers." External customers are the ones using Alibaba services, but internal customers are our colleagues, partners, and collaborators, who also rely on the work we do to achieve their goals.

Dual-Track Evaluation: Values and Performance

In Alibaba's early days, managers were required to give performance reviews each quarter. It was a lot to take on in addition to our day-to-day work, but we believed it was crucial in our formative stages. The extra attention paid to employee feedback helped create our strong company culture. Now, as a more mature organization with a firmly established culture, reviews are conducted twice per year. But the same principles that shaped the early evaluations continue to be used.

Most importantly, the Alibaba values (which include customer first) are central to performance reviews. Evaluating a team's adherence to those values not only helps us understand the process that leads to fulfilling business goals, but it is also a method for determining how closely employees embrace the culture. We called these dual-track evaluations—performance reviews broken down by the business results and company values.

Business results, or KPIs, are usually quantifiable, so those conversations revolve around whether we feel staff members are working up to their potential, whether we're utilizing the best methods for maximizing results, and if there are some tasks that require assistance or support. This is, of course, standard at most companies, but it's also where a lot of companies end their performance evaluations. Not at Alibaba.

Our evaluations also take stock of how the employees' performance reflects the company's values. Managers seek to determine if an employee is manifesting the company's tenets. The performance reviewers explicitly cite the six values that guide their conversations with employees. For each value there is a subdefinition that breaks it down further into four points. This gives the assessments more structure. At the same time, an HR staffer provides a third-party perspective and can mediate if the two sides see a value differently.

Applying Core Values

When assessing adherence to and implementation of values, reviewers rely on Alibaba's New Six Values. As mentioned in Chapter 5, Alibaba's previous values known as the Six Vein Spirit Swords were customer first, teamwork, embrace change, integrity, passion, and commitment. In 2019, they were updated to the following:

1. Customers first, employees second, shareholders third.
2. Trust makes everything simple.
3. Change is the only constant.
4. Today's best performance is tomorrow's baseline.
5. If not now, when? If not me, who?
6. Live seriously, work happily.

In addition to these values, evaluators are encouraged to consider if employees "do the right thing" and also "do things right." These values must function in concert with one another. If you fail to do the right thing in the first place, your results are going to be questionable. In other words, the ends do not always justify the means.

This became clear to everyone in 2010 when the company's Black-list Scandal occurred. Executives discovered that some members of the sales force had boosted their sales numbers by bypassing the necessary security and accountability precautions when signing up new suppliers. Many of the Gold Supplier accounts signed up from a particular sales region in Putian, Fujian province, were actually fraudulent accounts posing as companies but being used to cheat buyers.

This violated at least two fundamental values that had guided Alibaba from the start—customer first and integrity. Buying products from "verified" yet unqualified sellers put customers at risk, and a platform lacking integrity would destroy the entire foundation of trust on which the marketplace was based. Once Alibaba's leadership learned of the practice, they fired more than one hundred sales team employees. But the sense of responsibility did not stop there. Both the CEO and COO of Alibaba.com also stepped down.

Some commentators questioned the responses, arguing that the actual sales that these fraudulent accounts represented was immaterial by regulatory standards; therefore the actions far exceeded any financial impact from the improper behavior. But those comments missed the point. Alibaba's leadership wanted to send a clear message to everyone that the company's values were to be taken seriously—they were not just for show—and violations would not be tolerated.

Even though the sales team in Fujian province recorded robust numbers, their method of pumping the accounts by signing up these dubious sellers harmed the company's reputation for honesty and reliable quality. Jack believed Alibaba was selling far more than just China Gold Supplier services. What they were actually selling was trust to all the businesses that used the platform. If Alibaba could not guarantee this trust was protected and maintained, then the costs now and in the future would be far greater than just a few fake sales accounts. The Fujian team neither did things right nor did they do the right thing.

Not all value judgments are so simple. In many ways, performance reviews are quantitative affairs: employees are at least partially judged on numerical items, like sales targets, or the growth of new app users. But there are important qualitative judgments that must be made, too.

How can we factor in something as subjective as "values"? And how can we make sure that evaluators are giving these factors equal weight across the company?

For many years, Alibaba adhered to a system that assigned numerical weight to an employee's value adherence. Performance reviews included a portion in which the evaluator ranked an employee's values commitment on a scale of 1 to 5, with 3.5 being a good rating and scores such as 3.75 and 4.0 given to those with exemplary achievements. Meanwhile, scores of 3.25 and below were grounds for a warning that the employee's work was not meeting expectations.

In 2014, the company simplified this system and switched to As, Bs, and Cs. Employees who earn As are exceptional role models, exemplifying company values in a way that inspires others. Bs strongly adhere to values overall but could still improve, and Cs exhibit gaps in their understanding of values and must actively improve their adherence to them. These scores play a key role in the employee's overall performance evaluation.

Wild Dogs, Bulls, and Rabbits

Savio Kwan, Alibaba's first COO, designed a framework to categorize employee types based on their performance and values: star, bull, dog, rabbit, and wild dog. Stars exemplify company values and achieve stellar performance. They are rare and exceptional employees and should be rewarded for their outstanding contributions. Bulls exhibit good values and strong performance and have the potential to someday become stars. Dogs have neither strong performance nor abide by company values. They are clear candidates for removal. Rabbits embrace company values but underperform. They likely have positive attitudes, which means they can usually be guided to improve their work contributions.

The most dangerous to the company are wild dogs. While they often perform well, they lack the right values, meaning their presence risks reinforcing bad work behavior in other employees. In the long run, their flagrant disregard for values can turn into a cancer that spreads throughout the organization and impairs its long-term sustainability.

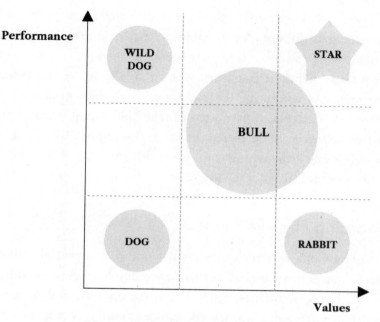

Classifying performance and values.

The sales team members guiding the Blacklist Scandal were almost certainly wild dogs. But when I think of wild dogs, another example stands out: the Mooncake Scandal. Each year, Alibaba celebrated the Chinese Mid-Autumn Festival by giving each of its employees a box of mooncakes, the holiday's traditional snack. These consist of a round crust filled with a tasty sweetened bean paste.

Alibaba's mooncakes are popular, so employees are allowed to purchase additional boxes through an internal online ordering service to give to friends and family. But in 2016, four Alibaba engineers took this employee benefit too far. Just to see if they could game the system, they devised a digital plug-in to hack the online ordering service and managed to obtain and hoard 124 boxes.

They were swiftly dismissed, but a robust conversation developed within the company's chat rooms questioning the response. Some wondered why what appeared to be little more than a prank was being treated as such a serious offense. Others pointed out that the trick demonstrated just how skilled the now-fired engineers were.

Though the engineers might have just thought it a practical joke, their managers did not. An Alibaba spokesperson said the hacking violated Alibaba's value of integrity. Particularly unnerving to the management was that these engineers were actually from the online security department. As wild dogs, these engineers were presenting a model of abusive behavior that, if permitted to take hold, could occur elsewhere in the company if not corrected. In short, the engineers had talent, but they were using it the wrong way—something that can become dangerous to an organization in the long run.

Rewarding High Performers

Rewarding employees for exceptional work is an essential component of performance evaluation. But there are multiple ways of providing the rewards: bonuses, increased salaries, stock options, and promotions. Alibaba has devised a framework for determining when to use each type of reward.

Bonuses recognize successes achieved in the previous year. Anyone who outperforms their annual goals should be rewarded in their annual bonus. In most cases, the standard bonus for solid performance was equivalent to three month's pay. Employees who performed exceptionally well received bonuses equal to four month's pay or more.

Two considerations influenced salary adjustments: market inflation (we needed to ensure that salaries kept up with what others were paying) or an upgrade in an employee's skills or job status.

Stock grants come in two forms. One is an onboarding stock grant, which represents an estimate of the value that the individual is expected to bring to the company in the future. The other is provided as an annual bonus in the form of stock rather than cash.

In the early days of the company, stock was given more generously because the company did not have a lot of cash but could create all the stock it wanted. Today, Alibaba is a mature company with significant revenues and a robust stock that the company does not want to dilute any more than necessary by issuing more shares. Since the stock is more valuable, it is granted sparingly. Generally, stock awards go only to those who have performed well and have demonstrated outstanding potential.

Job promotions are announced in May of every year. Employees can either be nominated by their supervisor for job upgrades or they can nominate themselves. Once the department head determines if a staff member qualifies, the employee appears before a committee, usually consisting of three more senior executives from other departments who understand the nominee's business unit.

These job promotion interviews are rigorous. The nominees must give a presentation on their accomplishments and what they have learned from their experiences. They must also demonstrate how the Alibaba values are reflected in their work and what they believe they can achieve moving forward. A spirited Q&A discussion follows, and then each panel member provides verbal and written assessments, which culminate in a vote. A majority is required to earn promotion.

The process is intensive, but the aim is to give the employee an opportunity to learn from the process and grow. It helps establish a clear vision of their path and the company's expectations. Having a panel of executives from other departments assess the candidate also ensures an impartial evaluation and fresh perspectives. The exercise results in a deeper understanding of Alibaba's business culture and values.

Identifying and Supporting Low Performers

It's every bit as important for supervisors to identify a team's low performers as well as its high performers. In many ways, a company's ability to support its low performers represents its commitment to employee success, something it should always demonstrate. The aim of the process should be working with those who need improvement, not just casting them aside.

Until recently, Alibaba's performance evaluation system worked on a forced curve. Team leaders were asked to identify the top performing 30 percent of their group, the middle 60 percent, and the lowest 10 percent. It should never come as a surprise when an employee is deemed a low performer. Employees failing to meet goals should be given notice during their midyear reviews or in periodic feedback sessions. This gives them time to adjust and improve before the in-depth annual evaluations.

Employees in this bottom rung are asked to put together a Performance Improvement Plan (PIP), which they create with the help of their evaluator. The PIP should be detailed and rigorous, which makes it an effective step-by-step road map to future growth. It's not punitive, but a guide. I've watched a number of my team members transform their performance as a result of their PIP process. Because they played a role in writing the PIP, they had a sense of ownership, which inspired them on a personal level to work toward a higher level of success.

For me, identifying the bottom 10 percent of each class was one of the more challenging aspects of the year-end reviews. I always hoped each employee was committed to working hard and succeeding. And I didn't want to discourage anyone or stigmatize them with this label. But over time, I saw this as a form of structured tough love that actually helped transform the individuals and their contributions.

Dean, a French-American colleague who was part of the first international leadership class, the AGLA program, is a good example. Dean was previously an entrepreneur in Shanghai. He had an easygoing charisma that naturally attracted followers. He was elected AGLA class leader and motivated his classmates.

These were excellent leadership qualities, but his supervisor was frustrated with his performance. She felt he lacked attention to detail. Dean had been managing a project focused on cultivating French online influencers to increase traffic and ultimately sales on Alibaba platforms from the region. Dean often captivated meetings with his big-picture ideas, but he shied away from delving into and analyzing the data. Put simply, he was more adept at evangelizing his ideas than in delivering results.

I narrowed in on this when I conducted Dean's review alongside his direct supervisor. Though she had tried to communicate this to Dean, he responded to her feedback with murky justifications. So in the review, we broke the bad news: Dean was in the bottom 10 percent of his class. We made it clear he needed to adopt a more data-driven approach and achieve real results.

Dean was visibly surprised, and I worried that he might start arguing. But after one long, awkward pause, he asked us to give him further details for how he might improve. As an entrepreneur, he'd been his company's ideas guy, he explained, and relied on his partners to implement projects. But now he started to realize he played a different role with more hands-on responsibilities. He needed to engage with the hard work of implementation.

Later, I followed up with Dean to ask him how he felt about the entire process. "To be honest, Brian," he said, "this isn't what I expected working at a company like Alibaba. Initially, I was a bit shocked to get this feedback, but then I was disappointed because I felt that I hadn't achieved the goals I had set for myself."

He admitted that he *had* considered arguing, but "I decided to listen instead and see if there was something else I could learn from the conversation. Be a sponge and take it in."

It was a watershed moment for Dean, who has since found lasting success at Alibaba. Dean went on to work as a business development executive at Tmall Global and spearheaded one of our first Africa projects, connecting Rwandan coffee growers with the Chinese consumer market as part of the electronic World Trade Platform (eWTP) Initiative, which grew out of the G20 summit in Hangzhou in 2016. The project was a soaring success, and to this day Dean not only leads a business development unit in Tmall Global but also has his own podcast and vlog, educating businesspeople abroad about how to reach Chinese consumers through e-commerce platforms.

Mapping Out a Time Line

All the elements of the performance reviews can be a lot to keep track of, so Alibaba also created a calendar where each step in the performance management circle has its place, all of it coordinated. Supervisors' time lines work in concert with their employees': both groups set goals in April, when managers communicate performance results and their bonus and salary adjustments. They spend the next few months coaching team members on their performance, and then review goals with them in October. They plan for the next full year in January and February, and in March they evaluate employees' overall performance (Alibaba's fiscal year starts April 1).

Of course, the performance management circle also relies on other players besides supervisors, including HR staff and upper management. Alibaba created a multitiered calendar with linked time lines for each participant in the performance-evaluation system.

Time Line by Different Roles

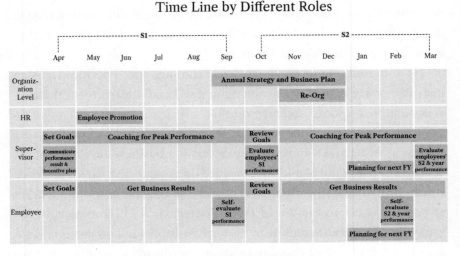

Performance management calendar.

The Employee Life Cycle

Performance evaluations take place on a twice-yearly basis, which can feel like regular check-ins within a lengthy career. But what happens when we step back and look at the broader picture?

A company's ability to look at a more encompassing long-term time line for their employees' careers has become increasingly important. Today, Gen Z is entering the workforce, and millennials are stepping into management roles. I've noticed that the younger employees often think more critically about the purpose and meaning of their careers that go beyond collecting material benefits. They're aware that their time at work makes up a significant portion of their overall lives and hope to spend it doing something they care about. How can business organizations respect that?

To attract and retain great talent, companies today must consider why people come to work for them and what they want out of it. They must ask themselves: How can we make a workplace the best possible experience? Why do people come to work for us in the first place, and why do they stay? In short, what is working at a company for?

Though many large companies are starting to think about these issues from the employees' perspective (particularly in the wake of COVID-19), Jack had been discussing this critical motivational factor for years. He believes, and often says, that good employees use their time within an organization to develop themselves. If Alibaba wants to be the best possible company, it needs to create opportunities for employees to grow—and not merely in superficial ways intended to boost KPIs.

As such, Alibaba considers its employees' overarching personal goals of equal importance to the smaller objectives related to their jobs. Jack believes Alibaba is a place to gather life experience, a training ground for real-life lessons. It's important to remember that a career is a journey, and an organization places everyone on a journey together. As we say at Alibaba, we are "well-meaning people who get together to do meaningful things."

We use the term *employee life cycle* when we talk about this bigger picture. If you think of an employee's tenure at a company as a lifespan, you can break it down into four phases:

1. Selection
2. Development
3. Retention
4. Rotation

The employee life cycle model ensures that we honor employees' personal growth and also respect that each employee has needs and aspirations.

Selecting the Right People

When we launched the Alibaba Global Leadership Academy, or AGLA, in 2016 to train a new generation of international leaders, we received close to three thousand applications for the thirty-two spots. When we organized recruiting events in places like New York or London, the auditoriums were packed with enthusiastic young people.

The candidates' qualifications were as impressive as the total number of applications. The applicants came from the world's most prestigious schools—Harvard, Stanford, Wharton, Oxford, Cambridge, Indian Institutes of Technology, University of Singapore. Their backgrounds spanned technology, banking, engineering, government, and even music. I asked myself how on earth were we going to filter through and select the right candidates from such a large number of highly qualified individuals? I was able to turn to Alibaba's framework for employee selection.

We consider the employee life cycle early in the hiring process. Alibaba selects people from the outset who share its outlook toward personal growth while at the company. To create teams of "well-meaning people doing meaningful things," managers need to hire people who will align with the organization's values and mission.

Viable candidates must demonstrate a sense of purpose larger than their own self-interest. Are there issues or causes they care about that have motivated them to take chances in their careers or even in their everyday lives? Have they committed to things that might have seemed irrational from the outside but were critically important to their own values? In our selection process, we look for those who have demonstrated that they will fight for what they believe in.

As we move further into the era of the fourth industrial revolution, knowledge workers must not only be competent in the technology they're working with but must also must possess the *intelligence* to be able to visualize how it can be applied to other areas, perhaps online or offline, and across sectors like commerce, entertainment, and health care. Employees' ability to visualize these new applications can help the company grow into and remain at the cutting edge of their industries.

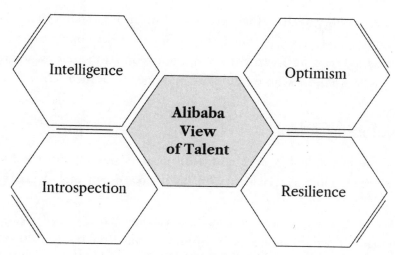

Criteria used for identifying talent.

As mentioned previously, early on at Alibaba we prided ourselves on being "simple and innocent." What we meant was that a bit of simplicity and innocence is helpful in an innovation-based business because it means the employees are unafraid of using their imaginations to develop fresh ideas never seen before. Those ideas might prove meaningful, but there's also a high chance of failure. It takes a healthy sense of *optimism*—or openness—to push on in spite of such setbacks.

Resilience is a close kin to optimism because failures are often the best instructors. We learned this over and over at Alibaba, given the potholes we encountered—the bankruptcy near miss in the early years, the quarantines during the SARS outbreak, the threat of eBay's entrance in the China market, the Blacklist Scandal, and the infamous "October Siege," when a group of disgruntled merchants staged spirited protests outside Alibaba's offices in response to the company's policy changes to elevate the quality of marketplace services. Alibaba seeks employees who are emboldened by the challenges they encounter.

The final attribute we look for is *introspection*, the ability to step back at regular intervals and reflect. Self-reflection is critical to addressing weaknesses and fortifying strengths. Alibaba's most respected leaders—Jack, Joe, Lucy, Daniel—possess this introspective quality, as well as an ability to share their personal realizations and insights. Their candor is a model for other staff, supporting their development. As I supervised my own staff over the years, I noticed that the ones who grew the most were those who were honest and open with themselves.

Developing Talent

Once the new talent is on board, team leaders should shift the focus to talent development, which Alibaba breaks down into a series of steps.

We hold a new employee orientation called Bai'a, or 100 years of Alibaba. For senior management hires, it's called Bai'hu, or 100 years of Hupan, a reference to the original apartment building where Alibaba was founded. In fact, the Bai'a training program was created from the moment we first codified our company mission, vision, and values. Jack and his founders realized just how important it would be to have a unified team committed to the same purpose. In addition to our culture training, Alibaba trained employees in operations, data management, marketing, and user experience design.

Sticking to Jack's favored *wuxia* theme, many of our training modules have names from the martial arts novels. Leadership seminars for front-line managers are called Xia Kexing, or the Hero's Walk. Midlevel leadership seminars are called Gong Shou Dao, or Way of Attack and Defense, the name of an original short kung fu film Jack once made with the famous martial arts actor Jet Li.

Meanwhile, the senior management leadership seminar is called the Tao of Alibaba. And Jack's Feng Qingyang class and Daniel's Xiaoyao Zi class are geared for cultivating the company's high-potential next generation leaders.

This ongoing emphasis on training and development—combined with the rigorous performance reviews—not only enhances employees' skills but also makes their time at work feel more enriching. For many, being empowered to learn and grow, taking inspiration from martial arts legends to fight for one's *own* cause, is what makes them want to stay with Alibaba.

Retaining Workers

One of the hallmarks of many of the executives who have been with Alibaba for a meaningful period of time is the diversity of roles they've played throughout their company tenure. For example, Sabrina Peng, the youngest Alibaba partner, joined the organization in 2000 and started out as a manager in the China customer service department. Next, she led the operations for the China and later the international B2B division and then cofounded AliExpress, one of the group's global retail marketplaces. Subsequently, she joined Alipay as general manager of its international operations and then became chief marketing officer of Ant Group. Today she serves as president of Ant Group's Social Good and Green Development Business Group, leading its social responsibility efforts.

In other companies, it might be unusual for an employee to bounce among so many different business roles and divisions, as Sabrina has. But it's standard practice for talent at Alibaba. Company policy states that when an employee works in a talent-based position for more than two years, he or she can transfer freely within the company. As a result, some of Alibaba's long-standing employees, despite staying at the company for more than a decade, have enjoyed remarkably varied and rewarding careers.

Those curious, hardworking, and talented employees will stick around if they feel their needs for personal growth are being met. In this way, career development and talent retention go hand in hand. This system rewards employees for devoting their talent to Alibaba by encouraging them to follow their passions. Performance reviews also serve as a great opportunity for employees to share which projects and opportunities excite them.

At Alibaba, managers take such employee interests seriously. The result is not only satisfied employees but also ones who approach their work with passion and drive, knowing that the rewards will include potentially even more interesting assignments in the future.

Rotating In and Out

The end of the employee life cycle, when things do not seem to be working out, is rotation, either within or outside of the company. The first principle of rotation comes from Jack Ma directly: train, remove, fire.

Sometimes employees just need additional coaching and training, which we can accomplish using the Performance Improvement Plan. Other times, they're better off being removed from one department and reassigned to another more aligned with their skills and interests. Firing only comes if all that has failed, as a last resort. There are also clear standards for acceptable behavior, which are reflected in the company Code of Conduct. Any actions that violate the code usually result in termination.

Otherwise, performance evaluations continue to be the standard for determining removal. Only after employees fail to improve on the problems identified in their PIP should they be asked to leave.

Performance Management: Why It Matters

At Alibaba, we make it our mission to foster employees' goals for themselves, for their teams, and for their overall career development, and we encourage employees to dream big. But dreams are harder to achieve in practical terms when they don't come with accountability, and that also makes a strong performance management framework crucial. Twice yearly performance reviews may sound tough, but they keep employees on track. In my own case, they pushed me to take ownership of my goals while teaching me self-reflection.

I did not completely understand Alibaba's performance management principles when I sat down for my review that April, in 2006, but once I engaged with the process, they fortified me. Like the sages in Jin Yong's novels, my superiors in the conference room offered guidance to help me sharpen my business and personal acumen as well as providing strategies for success.

Since then, I've come to appreciate how the performance management circle supports continuous reflection and self-improvement through disciplined goal setting, twice yearly evaluations, coaching, and rewarding. I benefited from remaining at Alibaba long enough to experience the employee life cycle—its selection, development, retention, and rotation stages—play out for myself and others.

Of course, in my previous jobs in the United States, I had learned about the importance of talent in driving performance. But I had never worked at a place that, particularly in its early days, cultivated talent with as much discipline, consistency, and thoughtfulness as Alibaba. By placing so much emphasis on the role of values, the performance conversations made the process rigorous in positive ways.

Though it took my moving to China to learn all this, Alibaba's performance evaluation culture was partially inspired by American businesses. Savio, who came from GE, introduced some of their frameworks into his methods. We also took into account ideas from scholars like Robert Greenleaf and Peter Senge, who have written that cultivating your team is like cultivating a garden. At the same time, Jack and the leadership team added their own elements, which provided distinctive Chinese features to our performance system. They sought to cultivate gardens just as lush and rich as Jin Yong's Peach Blossom Island during a springtime bloom.

9

Leadership

Leading Without Leading

At the beginning of my second year of business school, I was selected to give a speech on leadership to the incoming first-year class. Wharton was holding the convocation ceremony on Penn's campus in Irvine Auditorium, a historical red brick building with handsome ornate white moldings, and I was directed to deliver my remarks from a pulpit in front. I felt like I was about to deliver a sermon at this temple of capitalism.

The auditorium was filled with energy and expectations. The first-years buzzed with chatter. Their enthusiasm gave me a charge of adrenaline but also some nervousness. From the time I received the speaking invitation, I had been anxious, wondering what I could possibly tell these high achievers that they didn't already know.

So I had decided that, in talking about leadership, instead of issuing a string of feel-good platitudes, quoting Peter Drucker or Steve Jobs, I would take a nontraditional approach and encourage them to question some of the traditional values that most of them, no doubt, would be carrying with them. I began by reading a passage from Laozi, the great sage of Taoism, that essentially turned the tried-and-true notions of leadership on their head:

> The greatest type of ruler is one of whose existence the people are hardly aware. Next best is a leader who is loved and praised. Next comes the one who is feared. The worst is the one who is despised. When a leader doesn't trust the people, they will become untrustworthy. The best leader speaks little. He never speaks carelessly. He works without self-interest and leaves no trace. When the work is accomplished, the people say: "Amazing: we did it all by ourselves."

I felt quite satisfied with my stab at inspiration, but as I looked out over those future leaders, all I saw were blank stares. The auditorium, once filled with energy, was now awash in silence. Anxiety swept over me as I wondered if my audience was lost in thought over the profundity of my statement, or if they were simply baffled. I knew that for many, leadership implies dominance, control, fearlessness. Yet, that day I was trying to share a little of what I had learned on my own journey and my experiences with leadership in China, a quality, I had found, that works best when delivered with humility and sensitivity.

In giving my supposedly provocative talk that day, I knew what I wanted to say but I was not at all sure that they had heard me. Were those students ready to entertain an alternative path to achievement? Was learning something new part of their own path to leadership?

An Early Fascination

I had dabbled in leadership opportunities growing up through various roles in student government and extracurricular activities. At Wharton, I led a task force for the dean's advisory committee to evaluate the school's leadership programs, which was probably why the administration asked me to be the convocation leadership speaker. I wanted to convey a sense that leadership was more complex than some of the more common descriptions suggested.

Growing up, I had looked to recognized leaders as models as I wrestled with what I wanted to do in life and wondered whether I would one day assume some kind of leadership role. I followed the news about those seen as the titans of industry. Jack Welch was the country's most famous CEO at the time. He grew General Electric at a rapid pace, but he was also hard-nosed, and famous for slashing jobs without warning. Welch was successful at what he did and admired, but he seemed to play a zero-sum game.

I also turned to politics. The American public debated the characteristics of great leaders every election cycle, of course, and I carefully observed the leaders I worked with early in my career in the office of Willie Brown, the mayor of San Francisco. I was fascinated by the question of how an individual can instigate so much change.

I attended seminars on leadership while in business school and even enrolled in a course called "Power Lab" with the renowned Professor Kenwyn Smith at the University of Pennsylvania's Social Policy School, to get a feel for power and group dynamics. I was hoping to understand not *only* the theoretical framework of leadership but also the psychological dynamics of influence.

At the same time, I followed the remarkable transformation taking place in my hometown, Palo Alto, where the tech industry was booming and establishing a new kind of leadership profile. Many early tech founders and entrepreneurs exhibited a more approachable leadership model than CEOs in other industries and those from earlier generations. They were collaborative and accessible and embraced teamwork. Some of these leaders also portrayed themselves as committed to social change just as much as economic progress. But the question remained whether these individuals and their innovative leadership approaches were reaching the communities that needed it most.

I wasn't looking for answers to this particular question when I decided to join the scrappy start-up in Hangzhou. But when I returned to Alibaba after my time away in business school, I saw that Jack Ma's capacity to lead and his style had flourished. By then, the company had grown from a few dozen people to more than a thousand team members. They had moved from a single apartment to their own dedicated office building, and Jack was emerging as someone important—not just at Alibaba but across China as the company's profile rose.

Alibaba was no longer an aspiring digital platform company but was seen as a business pioneer with the power to effect real change in the Chinese markets, and Jack's leadership style was recognized as a key part of that impact.

Jack captivated audiences in his public appearances, displaying a powerful charisma, flavored with a bit of rebellious swagger. He has a larger-than-life presence and can command a room with virtually any audience, from business tycoons to heads of state. Yet within Alibaba, I often witnessed aspects of his leadership style that were altogether different from anyone else I'd observed or studied. He was empathetic, earnest, nurturing.

If anything, aspects of his leadership style on occasion resembled some of the principles described by Laozi in the quote I'd read in my Wharton speech, particularly those related to empowerment of others. Jack engaged thoughtfully with Alibaba employees and sought out those conversations. Perhaps these qualities came from his previous life as a teacher, but he seemed to have a natural knack for enabling those around him, letting people know he was listening to them and expressing his authority by encouraging others to take charge.

He was driving constant innovation and steady change at the company, but often instead of just asserting his decision-making power, he cultivated the leaders around him. He did so through his ability to collect information from diverse sources, compile the complex set of topics, digest them, and then formulate his ideas into a set of simple yet often farsighted statements for others to embrace and then put into action. Whether it be the company's mission and vision or its values, shaping and expressing these principles were among Jack's most important contributions in the early stages of the company's growth. These are the elements that provided the guiding light and the guardrails for subsequent management decisions.

Jack's leadership style also reflected a characteristic of the new digital era. Organizations and their information flows were becoming more complex, and business cycles were accelerating. In response, companies needed a more distributed decentralized structure to ensure they innovated at a sufficient pace. A key to strong leadership in that environment was knowing how to unleash the talent and abilities of the team.

As discussed in the previous chapters, a capable leader first defines a company's mission, vision, and values so that the employees know why they are there and what their purpose is. This sets the framework and direction for strategy planning and organizational structure, while also providing guidelines for selecting the appropriate talent and assessing performance. Savio Kwan, Alibaba's COO, used to say that a leader's first responsibility was establishing and communicating the mission to the team, and then selecting people smarter than himself to pursue the goals.

The Alibaba leadership model is built on empowerment, humility, mental dexterity, empathy, and, not least, self-knowledge. These attributes, when well organized and manifested in the culture, spur passion in others, forge bonds, inspire loyalty, and unite teams around common goals. In the remaining part of this chapter, we will explore how this leadership style fits into the Tao of Alibaba, but we'll also see the ways in which it is relevant beyond the company, and why it will become crucial in today's new digital economy.

The Key Leadership Quality: Empowering Others

At a 2010 meeting with newly promoted leaders, Jack explained a key philosophical principle. "If you don't want to be a leader, you don't want to manage a team, it's fine," he said. "But if you are determined to be a leader, whether you like it or not, I have to tell you—your job is to help others be successful."

As the new managers listened, he continued: "If you have seven employees under you, these seven employees' happiness, family life, and ability to buy a house and a car will be in your hands from that day on. But many have not thought about it this way."

This was vintage Jack and exemplified how he saw the role of the leader. Rather than digging right in with KPIs or explaining how the newly promoted managers could master their workflows most efficiently and achieve higher targets, he compelled them to think about the happiness of their employees. In his view, those considerations came ahead of traditional corporate goals.

At Davos in 2019, Jack spoke to members of the World Economic Forum's Young Global Leaders. When asked what qualities he looks for when bringing someone onto his team, he said he hires people smarter than he is. The ideal candidate for a job, he said, makes him think: "People like him, four, five years later, he could be my boss. I'd love to work for him."

In fact, most of Jack's early choices of lieutenants fit this approach. Joe Tsai, Jack's first CFO and now executive chairman, is a Yale-trained former New York tax attorney turned private equity investor. Few people were more qualified than Joe to take on this important job. Savio Kwan served as head of GE Medical China before joining as the company's first COO. And John Wu, Alibaba's first CTO, was previously the chief architect of Yahoo Inc.'s search engine and was awarded a patent for its core technology.

Of course, Jack's philosophy may have been unconventional, but it was not entirely new. In the 1970s, Robert K. Greenleaf introduced the idea of "servant leadership," a model in which an organization's leaders prioritize employees' personal growth. And Esther Wojcicki, a renowned educator (in fact, my former high school teacher) and one of the key influencers behind Google's work culture, promotes the leadership approach called TRICK (providing Trust, Respect, Independence, Collaboration, and Kindness).

"People generally assume that transforming companies from good to great requires larger-than-life leaders," wrote management guru Jim Collins in the *Harvard Business Review*. He conducted a five-year study of successful companies and found humility was a key tenet of what he termed Level 5 Leadership, a "paradoxical combination of traits [that] are catalysts for the statistically rare event of transforming a good company into a great one." A Level 5 leader, he wrote, is someone who "blends extreme personal humility with intense professional will."

In his 2015 book *Team of Teams: New Rules of Engagement for a Complex World*, retired four-star general Stanley A. McChrystal also describes an ongoing shift in leadership structures based on his experience commanding US Joint Special Operations Command (JSOC) in Iraq. McChrystal found that the US military, even with the aid of sophisticated technology, was struggling to fight Al Qaeda's decentralized network of cells. He realized the US military's traditional top-down, hierarchical structure was not working in that environment. To truly push back Al Qaeda, the military needed to innovate. He ended up devising a more resilient structure to maximize agility and improvisation. Rather than relying on the traditional linear chain of command, the new setup consisted of a network of smaller, interconnected groups—the "team of teams"—which was more adaptable to changing conditions in real time.

While Alibaba faced a very different sort of battlefield in the digital space, General McChrystal's book reminded me of the ways Alibaba changed the traditional leadership structures to confront the rapidly changing digital market landscape. Throughout the company, team leaders and employees are in constant dialogue—up, down, laterally, and diagonally.

There are strategic reasons for this, not just on the battlefield but in the conference room as well. To continually innovate, a modern "smart" company can no longer just filter orders down from the top but rather must share information in all directions to develop the most effective responses. This process hinges on the participation of all levels of the company. It's a leader's responsibility to make sure that communication barriers and silos are removed and collaboration is encouraged as an essential virtue.

Cultivating Humility

When I decided to leave the United States for China to work at Alibaba the second time, several years after graduating from Wharton and working in the United States, I felt more confident about my decision than I had before. The first time I went to work for Alibaba, I had little in the way of real management experience and no formal business education. But this time around, much had changed. I had an MBA and more experience. I was excited to be heading back to China as a more fully formed leader. I was certain I would have a greater impact.

What I found when I got there was that Alibaba had also changed. The company was practically unrecognizable from the operation I had left behind just a few years earlier. Before, I'd known the names of every person at the company. Now, in my role as a senior director, I was in charge of three departments with staff members in Hangzhou, Hong Kong, Switzerland, and California. I oversaw our international paid product, called Trust Pass, our global marketing and business development efforts, and, finally, a Japanese-language website. And no, I didn't speak Japanese.

Though I was certainly more qualified on paper than when I left, I developed a case of imposter syndrome: How, I wondered, could I manage three business lines with employees spread across three continents? How would I guide them without sufficient experience myself in making these sorts of decisions? What if I didn't have good answers to the myriad business challenges the markets would be throwing my way?

The transition from individual contributor to team leader is especially difficult at a growing company whose philosophy is "embrace change." When I'd left the United States for China to take up this new role, a friend had given me a book about "your first one hundred days on the job" as a departure gift, and a couple weeks into my new role, I devoured it. But I had to laugh. Wasn't I supposed to know these things already?

Managing Alibaba's international marketing operations at that time was not for the faint of heart. I was told to apply what was euphemistically called a zero-budget marketing strategy, which simply meant I couldn't spend money on advertising let alone travel out of Hangzhou to meet potential overseas partners, most of whom had no idea where Hangzhou was.

Alibaba had a fraction of its current resources, and yet my mandate required that I compete with established international e-commerce players. I was, in short, being asked to deliver miracles, and I was convinced I would fail, quite visibly.

As frustrated as I was at the outset, I came to understand how this challenge, while exceedingly difficult, was an opportunity for growth. Jack's decision to assign me this position indicated his trust in me, and I realized many of my fellow team leaders were in similar positions. Jack had entrusted us with serious responsibilities, and none of us wanted to let him down. He had been clear about what he wanted us to achieve and communicated how our achievements would fit into the company's shared future. We were motivated to do everything we could to prove that his trust was well placed.

Over the years, there were many times I landed short of our ambitious goals, but in this case our marketing team actually pulled off a miracle. Thanks to our US leader, Annie Xu, and the local team members' creativity and hard work, we managed to make meaningful inroads in penetrating international markets and earning Alibaba favorable notice at international trade shows. Our zero budget marketing ended up making us stand out. With no resources, our means of communicating with customers had to be clever and attention grabbing. We certainly never ran a typical agency-style campaign.

My early years as a team leader ingrained in me another one of Jack's principles: humility. It's easy to come out of business school believing you possess special knowledge, that you have something the world needs. But really, it's the world that has something to offer you, I found repeatedly, provided that you listen well and are open to ideas from your team and the markets themselves. That year, my team members showed me how to achieve the impossible. As the Alibaba leader Lucy Peng said, even in successful times, "the honor belongs to the team, while the responsibility is on you."

Becoming Mentally Ambidextrous

In more recent years, I've given presentations on Alibaba's leadership philosophy to emerging leaders both inside and outside the company, sharing Jack's insights and style as well as some of the hard lessons I had to learn. On page 205 is a graphic I often use to highlight some of the many constituencies a leader must keep in mind. These include clients, supervisors, colleagues, and subordinates. But I also emphasize that leaders must remain mindful of the company's larger societal goals, as well as its ecological partners.

That concept stems from our reformulated strategy in 2010, when Alibaba redefined itself as not just a platform but also as an ecosystem of platforms and partners. The shift recognized that Alibaba no longer thought of itself as a one-dimensional standalone business but as a collaborator and partner to numerous other businesses, not least the merchants relying on our platform for income. Alibaba itself had become a series of interconnected networks.

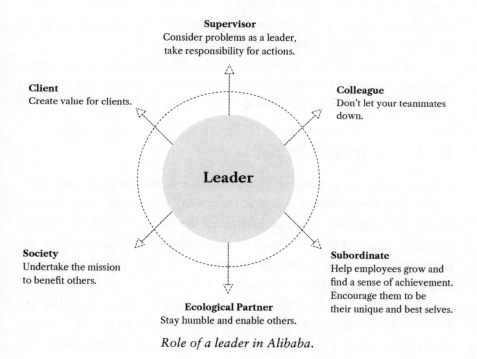

Supervisor
Consider problems as a leader, take responsibility for actions.

Client
Create value for clients.

Colleague
Don't let your teammates down.

Leader

Society
Undertake the mission to benefit others.

Subordinate
Help employees grow and find a sense of achievement. Encourage them to be their unique and best selves.

Ecological Partner
Stay humble and enable others.

Role of a leader in Alibaba.

This mental agility requires ambidextrous leaders. While it can seem overwhelming, the ability to be ambidextrous is one of the most important traits an aspiring leader can develop. Leaders must learn to balance their dream mission with the business reality, a kind of yin and yang.

Structurally speaking, a team leader should have a mind for over-arching business objectives while thinking like an HR leader about personnel and the individual team members' needs. That means a manager should never let the pursuit of KPIs overshadow attentiveness to employees' growth and success, nor should he or she let these concerns obstruct achievement of broader objectives.

Some might regard these differing priorities as contradictory, but leading effectively in a fast-paced, ever-changing company is, by necessity, about navigating paradoxes, oppositions, and parts of the strategic dialectic and finding equilibrium among competing interests. This ability to hold all these considerations in mind at once is what separates a leader from a manager. Leaders have followers, and managers have subordinates. If you think about it, one of those is far more difficult but far more inspiring and rewarding than the other.

"This is why we say 'ambidextrous,'" Lucy Peng once said in a presentation to newly promoted leaders. "A first-class leader should not only do well in business but also manage people well."

The Importance of Empathy

In the fall of 2017, business leaders from around the world gathered at the Bloomberg Global Business Forum in New York. Jack Ma was invited to speak, but when he addressed the forum he did not deliver the usual paean to the challenges of navigating a global economy or responding to the needs of the digital economy in the accepted corporate-speak. He was Jack, and he spoke from the heart.

"If you want to be respected, you need LQ," he said, using an acronym that meant nothing to the assembled corporate leaders and journalists. "What is LQ? It's the quotient of love, which machines will never have."

Everyone is familiar with IQ, an individual's intelligence quotient, and most now know of EQ, the more recently formulated idea of an emotional intelligence quotient. People debate whether IQ or EQ, hard or soft capabilities, are more important for success. But Jack was introducing a different capability for understanding and leading in the tech-innovation era.

"A machine does not have a heart, a machine does not have a soul, and a machine does not have a belief," he said. "Human beings have the souls, have the belief, have the value. We are creative; we are showing that we can control the machines." Digital tools are wonderful, he was saying, but they are only as wonderful in action as the human hearts guiding them, and that is where their value resides.

It wasn't the first time Jack had used the term—he first spoke about it publicly at an APEC CEO summit in Manila in 2015—but it was his first time sharing it with a Western audience, and it garnered a great deal of English-language press coverage, some of it mystified. He was asked about it again at the World Economic Forum in 2018, where, this time, he clarified the relationship between IQ, EQ, and LQ: "To gain success, a person needs high EQ. If you don't want to lose quickly, you will need a high IQ, and if you want to be respected, you need high LQ—the IQ of love."

LQ is a hallmark of Alibaba leadership. In 2021, Daniel Zhang, by then CEO, said: "Being a good leader is not just about technical competency. That's the very basic requirement. On top of that, it's about love and compassion. Do you really care about the people you work with?" Having a big heart toward others is the bedrock of LQ, which is why it's important at every level of Alibaba leadership. And it's why Alibaba leaders are asked to think about society and ecological partners, and why they're encouraged to think of their team members as much as company profit-making.

This compassion is baked into Alibaba's DNA. You can see LQ in Alibaba's mantra of *customer first, employee second, shareholder third*— the love quotient means caring for people beyond one's own immediate circle, acknowledging that these are people with desires and needs. It's also present in Alibaba's mission to make it easy to do business anywhere, which ultimately aims to reach and empower smaller merchants, workers in rural areas, and previously marginalized communities.

A great example of how LQ plays out at Alibaba is in the payment app Alipay. Though people in China use the app today to pay for everything from lattes to laptops, it started out as a way to help small businesses, often ones that were overlooked by large banks and that lacked access to financial services. The product reflected that helping the underserved was built into the company's DNA from the beginning and it advanced that priority.

Eric Jing, executive chairman at Ant Group, Alibaba's fintech spin-off, likes to talk about "financial inclusion," the work of providing marginalized people with all the standard financial tools, enabling them to conduct business, access microloans, and compete with larger players on a more equitable playing field without having to bear excessive costs or impediments.

"Financial inclusion is a key enabler to reducing extreme poverty and boosting shared prosperity, yet many individuals have been excluded from the traditional financial-services system," Eric wrote in an opinion piece in the *Wall Street Journal*. Alipay's mission is to create a new kind of financial-services system that is fundamentally more welcoming and encompassing of everyone's needs.

In many ways, Alipay has already succeeded. The app's users include everyone from businesspeople in Shanghai to agricultural workers in Gansu province. And even among its users who were never excluded from the traditional banking system, Alipay inspires compassion on their part; the app frequently encourages users to donate to relief funds in the wake of tragedies, natural disasters, and social need.

Many Western business leaders and the media struggled to understand Alibaba's inclusive philosophy at first. Analysts often dismissed Jack's mission and vision statements as slogans, mere corporate window dressing. They responded to Jack's pronouncements skeptically, which is understandable given that plenty of corporate leaders announce grand visions they never follow through on. But over the years, Jack's unwavering commitment to Alibaba's mission yielded real results and impact in China, proving that he firmly believed in the LQ values and that they worked. Alibaba employees believe in LQ values, too.

The concept of "shareholder third" was especially jarring for many business leaders, but Jack had articulated that hierarchy of interests years earlier, in 2007, during the run-up to the company's Hong Kong IPO. I recall attending a presentation at that time to an audience of investors, analysts, and other hard-headed capitalists, where Jack delivered a clear message.

"If you don't agree with our statement—customer first, employee second, and shareholders third—then *don't buy our stock*," he said onstage, in a tone that was almost defiant. Alibaba had no need, he suggested, for speculators buying up its shares one day, just to sell them the next.

A member of the audience called out, "I wish you had said that before I placed my order!"

Alibaba's track record and value were clear, so this was just an attempt to be cute, a meaningless provocation, probably from someone at one of the city's big banks. Jack was unruffled. "It's not too late to sell your stake," he said.

Four years later in September 2014, Jack took a softer tone when he issued a three-page letter to prospective investors weeks before Alibaba went public in New York. There was an important point he wished to clarify.

"I have said on numerous occasions that we will put 'customers first, employees second, and shareholders third.' I can see that investors who hear this for the first time may find it a bit hard to understand," he wrote. But he held firm on this core belief. "Let me be clear," he wrote, "as fiduciaries of the company, we believe that the only way for Alibaba to create long-term value for shareholders is to create sustainable value for customers. So customers must come first."

Employees, he went on, came next because "without talented, happy, diligent and passionately committed employees," it would be impossible to have satisfied customers. In fact, the glory belongs to the team because they are the ones doing all the work.

When he addressed Alibaba's investors, he spelled out exactly what Alibaba stood against. "Our company will *not* make decisions based on short-term revenues or profits," he wrote. Instead, "our people, capital, technology and resources will be utilized to safeguard the sustainable development and growth of the Alibaba ecosystem." He closed: "We welcome investors with the same long-term mindset."

Many in the West are waking up to the promise of these principles, as well as the responsibilities they create. In 2019, the Business Roundtable, an association of roughly two hundred CEOs from the top American corporations, issued a landmark statement upending decades of fidelity to a profit-first business model. After long endorsing the view that "the paramount duty of management and of boards of directors is to the corporation's stockholders"—a priority that reflected the decades-long dominance of the Chicago School of economic thinking, articulated and championed by Milton Friedman—the organization's new charter explicitly called for corporate governance to focus on a wider net of stakeholders.

When I saw that important turnaround, I welcomed the statement and hoped that the roundtable's sentiments would quickly be translated into action. At the same time, I was struck by how Jack's instincts had been well ahead of his time.

In its statement, the Business Roundtable redefined what it believed to be a corporation's basic purpose. The new approach urged that corporations be sensitive to and create value for "all stakeholders," not just those who owned shares. In a bulleted list, it went on to identify those stakeholders, which included customers, employees, suppliers, communities, and shareholders, in that order. The statement promised Business Roundtable companies would exceed customer expectations; invest in employees with fair compensation and training while fostering diversity and inclusion; deal "fairly and ethically" with suppliers, both large and small; and support the communities in which they worked by protecting the environment and embracing sustainable business practices. They articulated a vision of corporations playing proactive, positive roles in society.

A few months after the Business Roundtable released its statement, the World Economic Forum released its "Davos Manifesto." This, too, redefined the purpose of a company. In its very first sentence, it said a company should engage "all its stakeholders" and then went on to list some of them—employees, customers, suppliers, local communities, and society at large.

In an article about the manifesto, the World Economic Forum's founder, Klaus Schwab—who had been a strong proponent of this stakeholder capitalism for decades—wrote that he hoped corporations would understand that they are "major stakeholders in our common future" and that they would "also work with other stakeholders to improve the state of the world in which they are operating."

This represented a historic pivot away from Friedman's ideological definition of a corporation. Alibaba had long understood and had been a pioneer in putting this philosophy into action, as reflected in Jack's formulation of the LQ principle. This is at the core of Alibaba's structure as a business ecosystem.

Gaining Self-Knowledge

Savio once told me that a leader must answer the question Who am I? before he or she can become successful. Regardless of what leadership style one may have, self-awareness enables a leader to know how to best use his or her abilities to be most effective. Knowing yourself will help you decide what to do or what *not* to do. If you are not aware of your shortcomings or faults, you may end up like a bull in a china shop. If you are aware, then you know simply to avoid the china shop.

Though they're no longer part of the formal evaluation criteria, a few leadership principles left an indelible impression on me from Alibaba's early management review system. These principles provided valuable opportunities for self-reflection and pushed my former colleagues and me to recognize how our leadership styles were shaping our teams and how we could improve them.

The first principle is *wen weidao* (闻味道). It literally means "smell the fragrance" or, in more concrete terms, it's an individual's ability to read the room. Good team leaders should be able to sense their team's morale and energy level when they enter the office. Are the team members happy? Motivated? Or are they feeling lethargic? A leader's ability to sense these emotions can help him or her change the atmosphere, particularly if past decisions have contributed to a sour mood.

The second principle is *zhao jingzi* (照镜子), or "looking in the mirror." This goes hand in hand with reading a room because a team's spirit is often a reflection of how team members have been led. When team enthusiasm wanes, a leader needs to be able to look within to determine how his or her own actions may have caused the situation. Part of being a leader is confronting reality, even if that reality is unflattering.

The third principle is *jiu toufa* (揪头发), which can be translated as "pulling hair." *Jiu toufa* is about stretching yourself, breaking out of your comfort zone, making an extra effort to achieve better results. The implication behind *jiu toufa* is that simply meeting expectations is not enough; top leaders should find ways to exceed them.

Receiving or even giving performance reviews can be stressful. Much is at stake, including morale and motivation. But when framed by these three principles, the conversations can be quite rewarding. The opportunity to reflect with open eyes on your true impact on a team and also to assess your own personal growth is rare in managing the day-to-day challenges of work. This set of criteria went well beyond KPIs and enabled us to tap into our intuition when assessing our teams, honestly look at ourselves through the lens of those we led, and, hopefully, "pull our hair," or go beyond our usual limits. Applying these principles unquestionably helped me and my colleagues achieve breakthroughs we might not have imagined possible and to develop as leaders.

The Nine Axes and the Path to Success

LQ and a company's leadership philosophy are important concepts, but they are not of much use unless leaders are skillful at putting the ideas into action. To facilitate this, Alibaba adheres to a set of methodologies it calls the *jiu banfu* (九板斧), or "Nine Axes." Each of the three levels of management has three of those nine axes, which distill the relevant leadership principles as plans for action. While much of what I described above are behavioral principles, the nine axes are tools to govern and manage behavior.

The nine axes, like many terms Alibaba uses to describe the tenets of its management philosophy, derive from Chinese mythology. According to folklore, the Tang dynasty general Cheng Yaojin was known for seizing control of the direction of battles with his fearsome axe-handling powers. The expression *nine axes* refers to special skills that can be used in battle, and a business leader with *san banfu* (三板斧) is someone who has mastered them. The Nine Axes empower a leader to guide the team to success, even in times of intense difficulty.

The graph on page 213 shows the three management levels—frontline management, middle management, and top management—and the skills that are applied to guide them.

Development Path of Alibaba Management
Nine Axes

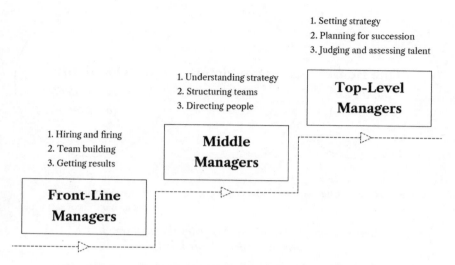

1. Setting strategy
2. Planning for succession
3. Judging and assessing talent

1. Understanding strategy
2. Structuring teams
3. Directing people

Top-Level Managers

1. Hiring and firing
2. Team building
3. Getting results

Middle Managers

Front-Line Managers

Development path of Alibaba management (the Nine Axes).

Front-line managers have the most concrete objectives: they must hire and fire, build teams, and get results. While these skills might seem basic to those familiar with management theory, a new leader's ability to master them often has the greatest impact on a team's ability to succeed. Hiring is an art, requiring skill in assessing a candidate's hard skills and his or her fit with the company's values and culture. The ability to identify and recruit potential talent has major ramifications for the team dynamic as the company grows.

Leadership is also key in creating a sense of unity for the team. (See diagram on next page.) Here Alibaba employs a particular approach: one heart (create an environment to build trust among the team), one picture (inspire the team to set a common goal), and one battle (develop your team talents through driving the business).

Compared to front-line managers, middle managers think more about the bigger strategic picture. They must understand strategy, structure their teams, and direct people. Understanding strategy requires

Build the Team

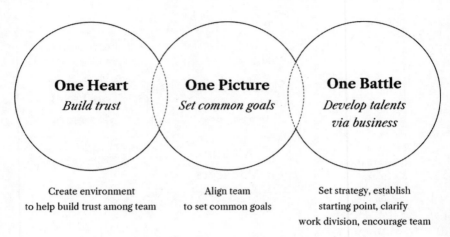

One Heart	One Picture	One Battle
Build trust	*Set common goals*	*Develop talents via business*
Create environment to help build trust among team	Align team to set common goals	Set strategy, establish starting point, clarify work division, encourage team

Create a sense of unity for the team through one heart, one picture, and one battle.

that they consider the why and the how of what they're doing. Team structuring necessitates optimizing resource allocation and directing people, meaning the leader must provide the conditions, guidance, and resources that enable others to act.

Top managers, meanwhile, need to shape overall strategies, plan succession, and use their judgment to gauge proper courses of action. With a good strategy defined, top managers must then facilitate the implementation, iterations, and experimentation with the plan in order to continue the process of refining and optimizing it.

Judgment is a key component of these roles, as well, and requires talent building on the team level, knowing which teams to employ for each task and how best to equip them. And that judgment should also apply to succession planning. In his study, Jim Collins notes that Level 5 leaders generally pick better successors. Because they embrace their company's overall mission and goals, they're more likely to concern themselves with how it will continue after their departure.

Leadership and the New Digital Paradigm

When I delivered my talk on leadership years ago to the incoming MBA class at Wharton, I had wanted to conclude with a personal anecdote that was meaningful to me, but I held back because the audience's reactions had been so muted. I wondered if I'd missed my chance to connect with them in some kind of personal way. Their applause was polite.

Later, after the orientation, one student approached and said she wanted to thank me for sharing the Laozi quote because it had really resonated with her. Then, a bit later, the head of our leadership development program at Wharton, Evan Wittenberg, came up and gave me a big hug. He said that, yes, the Laozi philosophy was different from what everyone at the school expected and was used to hearing but that very aspect made it exactly what the students needed to hear. It offered a provocative perspective, opening the door to fresh ideas.

Over the course of that week, many others I ran into on campus made a point of telling me how much they'd appreciated the remarks. Admittedly, it wasn't the roaring applause I had fantasized about, but I took it to mean they heard me, that I had reached them.

Now, I often share that same passage when I speak to leaders or prospective leaders on behalf of Alibaba. It's remarkable that such ancient thinking can be so timeless and relevant today. Laozi's quote is about the power of a more quietly encouraging leader and the importance of placing the honor of achievement and success on one's team. But it's also emblematic of the ways in which the Tao of Alibaba merges Eastern philosophy with Western management methods.

Two concepts that may seem contradictory can, in fact, exist in harmony.

Part III

The New Digital Frontier

10

Inclusive Development

From Hickory Nuts to a New Development Paradigm

More than once, when Jack was explaining his strong belief in the primacy of learning through experience, he would return to the story about his first trip outside of China, to Australia. As a teenager eager to improve his English, he relates how he once met an Australian tourist in Hangzhou and struck up a friendship. They became pen pals and, over a period of years, shared bits and pieces of their lives. When Jack was twenty, the man and his family invited Jack for a visit, and he leapt at the opportunity.

It was a pivotal moment in Jack's life. It expanded his horizon and gave him perspectives that were drastically different than ones that he was accustomed to. It was an eye-opening opportunity for him to see how economy and trade function in a Western society.

It was disorienting too, but it forced him into what has become a lifelong habit—challenging assumptions and placing trust in learning through direct experience, what he sees with his own eyes rather than textbook concepts or commonly held notions. At its root, the story, as Jack tells it, is about self-reliance and empowerment, providing confidence and clarity about the world and how it works.

That aspect of Jack's philosophy came back to me when I myself came to a turning point and sought him out for guidance. To some degree, my life path has been like a maze with numerous course corrections as I navigated back and forth between the United States and China, but in 2012, and then again in 2014, I again faced uncertainty, and Jack's advice—to find truth through lived experience, question assumptions, and seek confidence and clarity through real-world encounters—helped me.

That guidance would inform the next stage in my journey—and a significant new initiative for Alibaba as it sought to define its role on the global stage as more than just an e-commerce enterprise focused on maximizing profits.

Challenging Assumptions and Trusting Experience

Embracing the primacy of lessons learned through experience had been a guiding principle for me going back years. In college, I chose to write my undergraduate thesis about the experiential philosophy behind Henry David Thoreau's *Walden*. Living off the land in a cabin he built himself, Thoreau described the wisdom he gleaned from his surroundings and how they nurtured his vision of the path to the best life. He saw it as a form of honesty and sincerity. "I went to the woods because I wished to live deliberately, to front only the essential facts of life, and see if I could not learn what it had to teach, and not, when I came to die, discover that I had not lived," he wrote.

As a young man wrestling with the best way to pursue a career that embraced service and social contributions, I tried out a number of different avenues so that I could taste that life lived deliberately. That openness had led me to China, a detour to Wharton, and work at McGraw-Hill back in the United States, and then to China and Alibaba again.

My adventurousness was behind two important decisions I made in 2012. The first was to leave Alibaba a second time. The second was to undertake a new experience, running a marketing and media start-up in Hong Kong.

Behind the decisions was a feeling that my work up to that time at Alibaba, while rewarding, had run its course, that I had plateaued. My team and I had made a valiant effort to develop a new international paid product called Global Gold Supplier, a premium service that authenticated and verified international suppliers (particularly SMEs), and link them up with the world's buyers. We had set up a new operation in India and established partnerships in markets like Korea, Vietnam, Turkey, and many other countries. The initiatives had given us an expanded global footprint.

But I was unsettled. Despite my ongoing fascination with the work, my hectic travel schedule was causing burnout and fatigue, and I felt at this stage in life it was time to try out my own vision for a change. An old friend happened to reach out and told me about a Hong Kong–based social network platform that was raising funds and looking for a CEO. I leapt at the chance.

The business involved creating and overseeing an online community for Asian celebrities and artists to connect with fans and to develop marketing and branding opportunities. That aligned with an idea I had been toying with for some time. Some of the more advanced Asian manufacturers had attempted to build their own brands to add value to their product lines, no longer satisfied to just produce goods for Western brands at relatively low cost.

I believed that many of these manufacturers were ready to break out, but they needed help in creating strong brand identities. A potential solution was to build connections with known celebrities and leverage their popularity. I felt I was tackling the next hurdle in creating value for these entrepreneurs. I saw it as a potential renaissance of creativity for businesses in a part of the world that had developed a reputation as the home of copycats, not breakthrough brands.

Shortly after my arrival at the start-up, we celebrated our fifth anniversary with nice press coverage, and we signed a few new deals. But, as the months passed, the reality of trying to run a start-up and sustain growth on our limited capital turned into a grind. One day in Hong

Kong, during my regular hour-long commute on three subway lines to our offices, I started to think not just about the many hurdles to success but that this wasn't the best way to change the world. The number of people we were impacting was small, the process slow.

The team and I decided that the company might need to adjust its course. We looked for new ownership and were fortunate that we were able to sell the business to a listed Singapore-based company that had ample resources. It was a more respectable exit than entrepreneurs often have, but it left me searching for where I might go next.

Walkabout

Feeling adrift, I sought out a number of friends and mentors, from Silicon Valley to China, to help me plan my new chapter. Naturally, I found myself going back to Jack Ma, and as ever, he was a thoughtful source of guidance.

Jack was complimentary but said he felt I benefited from—meaning I needed—a leader looking over my shoulder to ensure I maintained a focus on my responsibilities. I needed a clear sense of purpose. He also took aim at my business school education, arguing that everything I had learned was, in the end, the opposite of what I needed to know as an entrepreneur. It had made me too risk averse, he said, focused more on KPIs than opportunities, especially ones that seemed improbable. Then he surprised me with his proposed remedy.

"Brian, the best thing for you to do now is go somewhere deep and remote in China's rural area," Jack said. "Find the poorest part of China and just travel through the region. When you find that place, stay for a few days. If you're comfortable there, then move on and find an even poorer area to stay. If you don't feel comfortable where you are, then force yourself to stay even longer. Then ask yourself, why do you feel uncomfortable? What is it about the environment that is causing you that discomfort? Once you go through this experience, chances are you will have more clarity about what you would like to do next."

I scratched my head at this one. Why on earth, I thought, would I want to take an antivacation and intentionally seek discomfort? As

I contemplated his exhortation that, in effect, I find myself by losing myself, it started to occur to me that the reboot he was urging might be worth the effort. This might be both a test and an opportunity. At least, I figured, I could clear my head, refresh my imagination, take stock.

It didn't take much research for me to determine that rural, mountainous Guizhou, in southwestern China, is one of the country's poorest provinces. It is also known for its natural beauty and villages of minority groups with rich cultures. I was fortunate to have met someone who was running an organization facilitating guided tours to more remote parts of China, and I was able to set up a trip that took me by train, car, and, at times, by foot to areas and villages off the beaten track.

I observed and digested life in Guizhou, but I was struck by a consistent reality, the dearth of working-age men and women. There were plenty of infants and young children in the villages I visited, but they were being cared for by the elderly, who also handled many daily tasks. Asking around, I discovered that the parents in these families had, in many instances, migrated to Chinese coastal cities to take higher-paying jobs in factories, sending money home. They returned just for a few holidays each year. These are known as "hollow villages" with "left behind children." The lack of economic opportunity in these remote areas was slowly draining away their vitality.

In one village, I sat in the town square one day to enjoy a performance of Chinese opera, a traditional art form with dramatic costumes, mythological characters, and doleful music. But as I scanned the scene, I realized that there were more artists performing on that stage than there were audience members, which were senior citizens clinging to small children. What kind of future did these Chinese villages have?

It was certainly not clear yet where this was leading. As Jack probably imagined, this was, at the least, a helpful antidote to the glitzy life I had been leading in Hong Kong and a return to the principles that had been so meaningful earlier in my career.

I wondered how I might help create the kinds of opportunities those rural villages were missing. How could I help entrepreneurs in remote areas so they could determine their own economic futures and not be forced to migrate and leave their children behind? In unlearning some of the business school principles I had absorbed, I was refreshing my sense of purpose.

And then I visited Shaji.

This Will Blow Your Mind

Not long after returning from Guizhou, I was visiting a former colleague at Alibaba, Chen Liang, who worked in the research department analyzing market trends. I mentioned my experiences in Guizhou and the sad reality of the villages, and he noted that, by contrast, Alibaba was learning of an unexpected rural phenomenon that some had started to refer to as "Taobao Villages."

Taobao was Alibaba's consumer shopping website, and he described these villages as often poor, rural hamlets where, on their own initiative, clusters of residents had formed e-shops, producing and selling on the Taobao platform a range of products that provided far more income than they had previously experienced. These entrepreneurial villagers, he explained, were building remarkable new levels of prosperity by local standards by developing small ecosystems of commerce, selling products such as farm produce, clothing, and furniture.

Further, these e-shops were creating a ripple effect, since other local residents were benefiting from their success by creating their own businesses to provide support services, such as packaging, delivery, and web design. It was a completely organic process, the result of inventive villagers often learning about Taobao and the opportunities by word of mouth.

The Alibaba researchers were fascinated. They began to study the success stories and were arranging visits for company staff and some academic researchers to learn more about how digital knowledge was spreading and boosting incomes through these enterprising clusters. It was a positive contrast to my depressing experiences in some of those hollow villages in Guizhou. Chen Liang mentioned a particularly robust cluster in Jiangsu province, in a village called Dongfeng within the Shaji township, that had a burgeoning digital ecosystem around do-it-yourself furniture production.

"Brian, I know you are thinking about your recent travels and of innovative ways to work with Alibaba these days, but there is something you really should see for yourself that will blow your mind," he said. "It's different from anything else we've seen. It's growing organically without us even knowing about it."

He added that he would be happy to get me on the next research visit being arranged. I jumped at this unexpected opportunity.

I was still living in Hong Kong at the time, 2013, so I flew to Nanjing in Jiangsu province. Then I took a bus for three hours and a taxi for another hour ride to Shaji township and made my way to Dongfeng. I got off at the village's gates and walked in to meet the team. This was nothing like the bucolic villages I had visited in Guizhou. The wide main avenue was dusty and relatively bare. There was a long row of traditional Chinese houses, each a cluster of rooms around a small open courtyard, mostly made of concrete and brick. Behind the houses was gray farmland that was largely lying fallow. This reflected the difficult efforts by the residents over the years to build their livelihoods.

My guide explained that for years the villagers had farmed, but the poor soil yielded modest crops. So some had switched to pig farming, but that, too, failed to boost incomes significantly. Finally, facing such impoverished conditions, villagers had turned to setting up small plastic recycling operations. In fact, it eventually became the destination of trash from the United States, Europe, and Japan. This refuse, shipped in containers, landed in a coastal port and then was transported to Shaji. There it was picked, sorted, cleaned, broken down, heated, and melted by local villagers; converted into reusable plastics and raw materials; and, finally, shipped elsewhere. The business had degraded the local environment over the years and earned Shaji township and its villages the unenviable reputation as "garbage villages."

The first indication I saw that the area's fate had taken a turn for the better were the Chinese characters painted on one of the walls in Dongfeng along the main road. It read, "Going away to work can't beat staying home and Taobao-ing."

I was introduced to Sun Han, who originated and led the transformation. He said that growing up in the Shaji area meant living with an unpleasant smell in the air. Two unsightly black ditches along the side of the road were filled with refuse from the recycling operations. The sky was always gray from the smoke.

He left the village for college and studied in Nanjing for two years but eventually decided to drop out to work. He took a variety of jobs, from stunt acting to an attempt to get into the wine business in Shanghai, but he did not find much success and drifted back home. He bought his first desktop computer and initially set up an e-shop on the Taobao platform to sell electronic devices and some locally made furniture based on his market research for potentially attractive product categories to sell. The business did not take off, so he paid a visit to Shanghai again and, during his trip, visited an IKEA store.

The store's inexpensive, do-it-yourself furniture kits intrigued him. He bought a few samples and returned home, where he showed the kits to a few local carpenters and hired them to design and produce DIY furniture for the Chinese market, which he offered online. The demand from Chinese consumers took him by surprise and produced, in the first month, more than 100,000 RMB in gross sales, the equivalent of about $15,000.

That was the spark that lit the way. Sun Han expanded his DIY furniture business, and sales became so brisk that *he helped other villagers* jump in and start similar online businesses. Some produced their own furniture kits, and over time, others started businesses to provide needed services, including printing, packaging, product design, photo studios, and website development.

As we strolled through Dongfeng together, I noticed lots of the small trucks commonly used for deliveries in Chinese cities parked along the street and buzzing about. There were piles of packaging materials in front of some houses or stacks of wood for furniture. This was not a typical rural village scene.

At Sun Han's family home, I saw that the traditional structure had been transformed into a mini logistics and supply operation. One bedroom had been turned into his makeshift office, what used to be a chicken coop and space for animals was used for storing wood and other production materials, and a third room was filled with packaging materials.

He explained to me that he had been happy to educate his neighbors in how to use the Taobao platform and provide advice on setting up e-commerce businesses. He, like the other villagers, was proud of this digital transformation and the widening circles of prosperity it was creating. The garbage village had transformed into China's number-one Taobao Village and real-world proof of the replicability of the digital ecosystems model.

How E-Commerce Platforms Remove Barriers to Market Access

It was, to say the least, an impressive display of the power of an idea, breaking through years of economic disappointment to raise the prospects of a one-time backwater, making it a demonstration of a new business model. And, as I had been told, it was a model sprouting up in other rural areas.

Walking with my hosts along the streets of Dongfeng, I noted that many of the homes had undergone or were undergoing a transition into some kind of business establishment, with the whine of the small commercial trucks providing the soundtrack of a new life for the residents. That was my epiphany: the Taobao Village paradigm was not only helping the residents of Dongfeng village; it was also potentially replicable and could be shared with others.

I left the area feeling energized. It was a reversal from my more contemplative, somewhat melancholy perspective after the trip through villages in Guizhou. That visit had impressed on me the depth and sadness of the problems of rural poverty in China. Shaji hinted at one possible solution. Alibaba, I realized, was really on to something that had the potential to help rebuild not just China's vast interior but, I thought, perhaps other emerging markets such as Africa or Southeast Asia, regions facing their own growth challenges.

It was a heady experience, and in a funny way, my exposure to a village ascending a ladder of 0s and 1s into prosperity was almost as jolting as Jack's first encounter with a truly modern industrial society in Australia. In my mind, I started to challenge assumptions about the long-standing futility of economic development efforts in low-income countries. It occurred to me that Alibaba might be able to disseminate and teach the e-commerce model that was proving an engine of economic transformation in rural China.

That engine was running strong. Today, the threshold for what is described as a true Taobao Village is an annual e-commerce transaction volume equivalent to $1.5 million and at least one hundred online e-shops, according to a World Bank study of the phenomenon. By that measure, the number of such sites shot up to 4,310 in 2019 from 20 in 2013, a remarkable statement about how this model for growing prosperity has spread organically in poor regions.

When I returned from Guizhou and then Shaji and sat down with Jack, I thanked him for what was a useful opportunity to brush away some of the unnecessary intellectual debris that had collected in my imagination. I acknowledged that, at first, the aim of the walkabout exercise was not completely clear but that, with time and a more relaxed daily pace, I had developed my own understanding of the yawning economic inequalities in Guizhou and ways the new digital model might be able to deliver inclusive prosperity.

As I related my experiences, Jack paused and then threw out a new proposal. He suggested that I come back to Alibaba—for a third time—now as his special assistant, bringing my new appreciation for the transformative power of digital ecosystems and Taobao Villages to the company's efforts at global expansion. He had a number of significant tasks he was working on—most importantly, Alibaba's planned US IPO in September 2014, the biggest in history up to that time—and he encouraged me to come back and help the company achieve these objectives.

Once I got over my surprise, I agreed and threw myself into what was, in effect, Alibaba 3.0. My initial responsibility was to help him prepare for the IPO, a landmark corporate event and a sprint that would conclude on the floor of the NYSE on a momentous day in September 2014. It was wonderful to take in the contrast between our earliest days on the couches in Jack's living room, struggling to figure out how we would succeed as a company, and the celebration of our arrival on the biggest corporate stage in the world as a Chinese, non–Silicon Valley global technology leader.

But I had already started to see that there was more to this story than an e-commerce success. Alibaba had built something that was changing once-poor Chinese villages, relying on the unique features of the digital economy and the Alibaba model. I was fascinated and motivated to see how we could further develop this breakthrough development paradigm.

Designing Curriculum for a New Development Approach

In bringing me back to Alibaba, Jack shared with me concerns he had about staying true to our values as we accelerated our growth outside of China. Yes, our successful IPO in New York was a great affirmation of what we had accomplished, but in strengthening our financial foundation and underwriting global expansion, it created two worries: first, how we could ensure that we maintained our culture and teach new leaders our guiding principles even as we expanded, and second, how we might be able to become more proactive in sharing our good fortune by teaching the digital ecosystem model to entrepreneurs in emerging markets as a new kind of development paradigm.

Following the IPO, I tackled the first of these issues in my next assignment, helping Alibaba build a talent pipeline strong enough to support the company's new global strategies. By scaling so rapidly outside of China, it was increasingly difficult to hire country leaders and other senior managers who were well versed in and completely embraced our systems and methods. Maintaining that culture was critical to our continued success, but our company culture also needed to be resilient. It had to evolve and embrace some of the practices and the thinking of the entrepreneurs and consumers we were encountering outside of China.

All of these issues and needs were turned into two programs that I would help lead, one to share the Alibaba model with our newly joined young global managers—we called that AGLA (mentioned briefly in Chapter 4)—and the other to work with budding entrepreneurs from emerging markets to provide training and support in deploying the Alibaba digital model, called Alibaba Global Initiatives (AGI). The two were similar in focusing on transmitting the basics of our operating model, but had different audiences and different impact. First came the founding of the Alibaba Global Leadership Academy (AGLA), where promising young people from around the world would spend a meaningful amount of time in training, gaining practical experience, before helping run overseas Alibaba business units.

In truth, I was initially reluctant to take these positions. I told Jack that it felt like yet another detour from the core discipline of building and running Alibaba businesses. But Jack persuaded me that training the next generation of Alibaba global leaders would leave a more significant legacy for the company and that this project should be led by someone with a deep understanding of the Tao of Alibaba.

In building AGLA, I paired up with Alibaba's international head of organization development training, Cai Song, and together we codified the company's core operating principles and mapped out a blueprint to explain Alibaba's emerging market growth paradigm.

In fact, through the creation of the AGLA curriculum, it became increasingly clear even to me that what we had to offer through our experience with Alibaba's digital platforms could be turned into a positive offshoot of the digital commerce revolution. The new model we were defining and teaching leveraged the capabilities and low barriers to entry of the digital economy and empowered entrepreneurs and rural communities. These groups had never been able to compete on an equal basis with enterprises in developed markets. The e-commerce revolution gave them more access and opportunities, and Alibaba was very much a contributor to this radical change that was taking place.

The program underscored the reality that education and training were core disciplines at Alibaba and the key to a new phase in the company's conceptual expansion.

The New Digital Frontier: Expanding Economic Opportunity in Developing Countries

Little by little, it was becoming clear that Alibaba and its approach to commerce might be offering a model for a potentially unprecedented transformation of the economic system. It represented a new frontier.

Digital platforms could empower populations that had been marginalized. And the demonstration of this concept was coming from the bottom up, not being handed down by the major multilateral development organizations, though they potentially had a significant role to play in supporting the spread of the new model.

One source of strength was the importance of data as a fuel and lubricant in this model. In traditional industrial economies, businesses are typically characterized by a zero-sum mentality—commodities like oil are scarce resources, and value creation comes from hoarding such resources and blocking out competitors. But in the new economy, data is an essential and a renewable, reusable resource that can be easily shared, consumed by multiple parties simultaneously, and—this is key—made more valuable as more applications are found.

Merchants use data for consumer insights, logistics planning, sourcing supplies, and new product development, while consumers use data in the form of product recommendations and cost comparisons, which then generates more data.

That is why one of the great innovations of the digital economy is that the objective should be "growing the pie" through network effects and collaboration, contrary to some of the old competitive principles taught in business schools. As a result, more value can be created for everyone with such ecosystems. The traditional barriers to entry are falling. Many of the entrepreneurs benefiting from the Taobao Villages, for example, had little more than a middle school education, little if any business experience, and extremely limited capital.

In distilling the essence of Alibaba's model of creating easily accessible digital ecosystems, in defining a central role for entrepreneurs in virtually any location and enhancing their capabilities with support systems, we were translating our success into a catalyst for inclusive growth in regions that had lagged.

Some experts have worried about the risks of emphasizing digital businesses, concerned that they might concentrate economic power with a tiny number of people who could just turn around and exploit poorer citizens, but I was clear that by far the bigger risk was in missing the opportunity altogether.

Populations around the globe that reside in the bottom 40 percent of income levels have largely remained poor despite the best efforts of multilateral aid organizations and national development strategies to break the shackles of poverty. The old models are poised to change, the faster the better. Digital technology stands out from previous economic breakthroughs in part because of the low threshold for adoption and the ease of replication.

When Jack first toured Africa in 2017, he observed that, in many places, what he was seeing was not so different from the poorer areas of China he had sought to transform with his platform in 1999. What many saw as Africa's shortcomings—heavy youth unemployment, weak infrastructure—were, in his view, opportunities to deploy digital platforms that would help entrepreneurs and small businesses leap over inadequate legacy systems and reach global markets. His views helped lead an important new phase in how Alibaba pursued its mission.

Supporting Access to Digital Commerce Through Training

Among the many ingredients critical to this model is support from public institutions to build the foundation. Government, for instance, should develop policies and regulations that encourage investments in innovation and that give entrepreneurs the freedom to build. Governments should also invest in the physical and digital infrastructure, especially in providing access to inexpensive broadband. Entrepreneurs should have support from venture capital and incubators, as well as business mentors. And educational institutions should be training each country's youth with relevant skills so they can participate in the new economy.

The explosive growth in internet use and e-commerce in China has shown the way forward.

By the end of 2021, e-commerce comprised more than 50 percent of all retail sales in China today, the first country in the world where online sales have surpassed offline sales. In rural areas, Alibaba produced $97.6 billion in e-commerce activity in 2018 and generated 6.83 million jobs, much of it through their Taobao Villages. The volume of e-commerce transactions in China totals more than the volumes in the United States and Europe combined. This reflects the way companies of all sizes and in nearly all parts of China have embraced e-commerce as an everyday tool for their work.

As I had observed, the Taobao Village phenomenon produced a strong multiplier effect. A 2019 World Bank study found that, of the 60,000 residents of Shaji town and its villages, 25,800 worked in e-commerce-related jobs. There were seventy-three logistics express companies, twenty-four photography companies, three e-commerce operation service companies, seventy raw material suppliers, thirty-six hardware accessories manufacturers, and fourteen accounting service companies.

The per capita GDP rocketed up in Shaji from $1,174 in 2008 to the equivalent of $5,029 in 2015. And in 2020 according to government data, Shaji's total e-commerce revenue reached 12.8 billion yuan (close to $2 billion). Much of that benefit was going to the digital workers. The per capita income of e-commerce households was 80 percent higher than other (non-e-commerce) households in the same communities, the study found.

There are many indications that this expansion and the multiplying levels of prosperity comprise a self-reinforcing process. Villagers, through their traditional connections with other family members and friends in nearby villages and towns, spread word of their success and the means to achieve it. Thus, neighboring villages are attracted to embrace the model.

This, too, has been studied. In a compelling economic analysis in 2019 by a group of leading economists under the auspices of the Luohan Academy, a research center initiated by Alibaba and dedicated to studying how digital technologies can solve societal problems, the data behind the spread of the Taobao Villages was compiled.

The paper, "Digital Technology and Inclusive Growth," found that "the rapid development of Taobao villages and towns illustrates the power of learning. Villagers spread the knowledge of e-commerce to their neighbors and neighboring villages through their connections. Neighboring families and villages are attracted by the success of their 'first-mover' neighbors to join the wave of e-commerce and are often involved in similar industries."

How New Fintech Technologies Can Transform Low-Income Communities

Another factor driving this transformation is digital financial tools, or fintech. Fintech tools can streamline and reduce the costs of engaging in all types of financial and commercial transactions, particularly for previously unbanked populations. They can also provide expanded access to credit for individuals and SMEs, especially in the form of microloans, by applying big data analytics instead of the cumbersome demand for collateral, the traditional basis for lending. Those barriers to credit have been significant barriers to economic advancement.

Fintech has started to change that, and China, which had seemed an unlikely case study in large-scale financial empowerment, has become a leader. The dissemination of the digital financial innovations, led by Alibaba, has brought about a new degree of inclusion in China. Ordinary daily transactions once involved wads of tattered and filthy paper currency, but China is now as close as any country in becoming a cashless society. That has helped narrow the economic gap between rural and urban populations and provided a key ingredient in the Taobao Village phenomenon.

Today, close to 90 percent of the urban population in China uses digital payments, as do 82 percent in rural areas. That has produced exceptional volumes of streamlined transactions. At the end of 2020, the number of mobile payment transactions reached a value equal to $67 trillion.

The impact has been particularly strong on SMEs, giving them access to tens of billions of dollars in credit. The fintech innovators are able to provide this credit in a fraction of the time traditionally required for loan approvals, without demands for collateral, by relying on large volumes of financial data, analyzed using sophisticated software and artificial intelligence. These are known as contactless loans, since they can be obtained by going online and providing data that the lenders are able to analyze instantly. Better, in many cases lenders can provide such credit with even greater certainty of repayment. This has been an especially important innovation during the pandemic, when face-to-face encounters were difficult if not impossible.

As an indication of the importance of this fuel for economic growth, SMEs, which create more than 80 percent of urban jobs in China, obtained $74 billion in fintech loans as of the end of 2020, according to the People's Bank of China.

Alibaba has put great emphasis on financial inclusion through a series of fintech innovations on its platforms. The explosive growth of those services is a testament to the enormous unmet demand for financial services and their power to accelerate economic growth. The services were initially bundled together as Alipay, which was created in 2004 and is China's largest payments platform. It was later folded into Ant Financial, which was spun off in 2014 as a separate company. It includes a lending facility that employs what it calls the 3-1-0 model. This service offers online applications that take three minutes to fill out, then one second to process digitally, while requiring zero people to actually fund the loan. Alipay now has more than 1.2 billion global users. And over the last six years, Ant Financial has provided credit to more than forty million SMEs, about half in rural areas.

The Luohan study cited research showing SMEs enjoyed a rate of growth, on average, of approximately 9.5 percent in the first month following such new borrowing, an enhanced level of expansion that persisted for at least six months. The credit reduced the volatility in sales of the borrowers, too, particularly during economic slowdowns, according to the Luohan research.

Another important building block in the digital growth model is something of unprecedented power and reach that you can put in your pocket, the smartphone. More than 60 percent of the population in low-income countries have access to mobile phones, an extraordinary level of market penetration. The digital paradigm that has worked so effectively within China would work just as well, we knew, in other emerging markets, spurred by this easy access to smartphones.

We had experienced this transformation at Alibaba, of course. In 2013, when Alipay overtook PayPal as the largest payment platform in the world, it was largely driven by the expansion of mobile payments. During this period, Alipay saw its desktop-mobile transaction ratio flip from 70–30 percent to 30–70 percent, a huge swing that added to our conviction that the convenience and power of mobile devices was of historic significance.

The New New Thing Made New Again

The digital e-commerce ecosystems not only spread prosperity and create jobs but are also something of an equalizer, providing greater opportunities for the disabled and women. Alibaba research has found that 50 percent of Taobao business owners in China are women, compared with 15 to 20 percent in the traditional economy. And by 2016 there were over 160,000 stores operated by those with disabilities, creating RMB 12.4 billion in total sales.

The fertile soil that nourishes the e-commerce seeds relies heavily on government infrastructure and collective support services. These include stable sources of electricity; functioning and properly maintained ports, roads, and bridges for moving physical goods efficiently; a cost-efficient postal service; and, of course, widely available internet connectivity as discussed in Chapter 2.

The Luohan paper described the new model as unshackling emerging markets from old constraints. "A nation's level of economic development is no longer decisive in determining its speed of technological penetration, and how quickly applications of a certain technology develop once it is adopted," it said. The paper stresses the three key factors that make the new digital revolution differ from previous economic transitions. First is the "low thresholds for the adoption of digital technology and penetration."

Platforms such as Taobao and Alipay were designed with simple interfaces to enhance easy use and access. The platforms have especially simplified interfaces to make them easy to use on mobile devices. Many of these functions can be accessed with voice commands, too, further opening the door to user access, particularly for the disabled and people with lower literacy levels.

In addition, Alibaba has devoted resources to training new users in utilizing digital tools. From the earliest days of its establishment, Alibaba offered programs to teach SMEs how to offer products online.

Beginning in 2006, Taobao started offering more than three thousand prerecorded training classes and twenty thousand live sessions a year, especially in poorer regions and remote areas. From 2015 to 2017 more than one million public sector officials from 765 poor counties in China took such courses.

A second critical element cited is "low user costs." The reductions in the costs of computer processing power and broadband access have accelerated these e-commerce developments. Mobile phones have, of course, further reduced the cost of this access in low-income countries and many rural regions previously too remote for such connections. The number of middlemen required to get products to consumers has declined, eliminating many expenses.

The third element of the revolution is "the non-rivalrous nature of digitized information." As explained earlier, one of the distinctive features of data is that it grows in value the more it is used, unlike many traditional industrial commodities, and it can be used, analyzed, and deployed in an endless number of locations at the same time. Not only is digitized information largely "non-rivalrous," but its wide dissemination helps support the ecosystems that support prosperity.

The Luohan paper concluded that all this has added up to China's digital economy making an economic advance of historic dimensions. "It is the largest integrated marketplace in human history, with unprecedented features: the largest number of participants—consumers and suppliers, especially MSMEs [micro, small, and medium enterprises] and startups; the largest variety of choices; and the greatest access to remote, previously undeveloped regions," the authors wrote. "It has fostered markets in which, for the first time, companies can start with small investments and serve customers thousands of kilometers away."

An Empire Built on Nuts

The Taobao model is flexible, and it can be built on a foundation of many different types of products. Take Bainiu, a now-famous Taobao Village that is about sixty miles outside Alibaba's hometown, Hangzhou. Historically, this was a prosperous town, resting at the site of a key bridge on an old Ming dynasty trade route that ran between Hangzhou and Anhui province. In more recent years, it was a quiet and not very prosperous agricultural town.

That began to change in 2007, when a local couple decided to create an online site to sell hickory nuts, which are similar to pecans. They called their e-shop Shanli Fuwa. They knew almost nothing about how to create an attractive website, so it was spare, but on day two after opening, they received a large order for 1.5 kilograms of the meat from peeled nuts from a buyer in Shanghai. From there, sales took off, taking the couple largely by surprise.

They created what became, in effect, a processing, packaging, and marketing operation. They bought the raw nuts from farmers in the area and, eventually, had to buy them from other regions in China and import them from as far away as Vietnam to keep up with demand for what had become a popular brand. Just as in Shaji, other villagers built e-shops along with widening rings of support operations to help this ecosystem flourish. Bainiu has become a darling of the Chinese media and is cited as a sterling example of the prosperity created by the Taobao Villages.

That was the vision I took to AGI, my next project at Alibaba—a training and knowledge transfer program for entrepreneurs from developing countries. The program helps entrepreneurs to build inclusive digital ecosystems and has changed the way they envision their ventures and do business. Our goal was to show that change at this scale is indeed possible and replicable. It was especially important to show that benefiting from this opportunity does not require a fancy college degree or a family pedigree. Ambition, a sense of purpose, curiosity, a willingness to learn and share—these have been the important elements of success in this new realm. In researching and teaching this model, I realized that I was achieving my dream from years earlier, of living life deliberately and advancing the collective good.

11

Going Global
The Students Become
the Teachers

AGI—Alibaba Global Initiatives—had a clear purpose: to impart digital knowledge to ambitious entrepreneurs from emerging markets and give them tools to build and utilize digital commerce platforms. Those were good intentions, but once we began to identify and engage with these enterprising pioneers, I quickly realized an entirely different dimension to our program. It was, in truth, a doorway to lands of exceptional talent and creativity where the participants introduced us to new levels of innovation and problem-solving, in spite of challenging infrastructure.

Carbon, a Nigerian fintech company cofounded by Chijioke Dozie, began by providing microloans and then expanded with a mobile app that offered a broad array of financial services to local consumers, banking the underbanked and unbanked. MAX, a Nigerian mobility start-up, led by cofounder Adetayo Bamiduro, morphed into a clean energy transport service that also focused on providing a wraparound financial safety net to its commercial drivers. And the Kenya-based venture Wasoko, the brainchild of founder Daniel Yu, is an innovative company providing tech-enabled supply chain solutions to Africa's long ignored but scrappy informal retailers.

The entrepreneurs behind these companies all became a part of our eFounders Fellowship community, graduates of the program that Jack Ma had championed and brought to life. Jack's insight was that the nations of Africa, Asia, and Latin America, historically burdened by poor infrastructure and big income gaps, stand to gain the most from digital ecosystems that spread prosperity and fuel inclusive economic development. The talent is unquestionably there—we have found over and over again—and given opportunity and tools, it has soared.

As I learned more about these AGI entrepreneurs and many others, I came to see the many dimensions of their achievements. They had discovered and embraced true values and purpose in their enterprises, and at the same time, they were instilling confidence and pride in the local communities through their success and example. They understood that they were creating a new kind of business model for their markets and demonstrating a new sort of mind-set for their local peers.

Carbon's founders, for example, not only empowered Africans by providing greater access to financial services, they also decided to publicly disclose the company's financial information to counter a lack of trust in the local business environment. As a result, the company has been recognized for its bold approach to transparency and strengthening trust among customers, partners, and investors. MAX's founder made it his company's mission not only to provide job opportunities for commercial drivers but also to create a platform to address their life needs, including skills training, health insurance, educational financial aid, and banking services. In response to the fact that more than two-thirds of the nonagricultural jobs in Sub-Saharan Africa are in the informal economy, Wasoko's founder has focused on utilizing his business model to help mom-and-pop shop owners reduce costs and increase margins, a game changer for many owners of roadside kiosks and shops.

These companies and many other eFounders ventures have not only flourished in their local markets; many have also succeeded in raising additional capital, working their way toward unicorn status. Our educational and support programs offered training for these entrepreneurs, helping them understand and exploit digital ecosystems, but Alibaba often enjoyed little benefit beyond seeing its model flourish. A few of these companies, such as Flutterwave, in Nigeria, ending up raising capital from entities that *competed* with Alibaba's businesses. And a number of participants in the AGI program ended up collaborating or partnering with Alibaba competitors.

From the standpoint of the traditional zero-sum business model, the dominant business mind-set for more than a century, I should have been disappointed that Alibaba passed on opportunities to invest in some of these companies. They were, after all, rising stars in Africa, and an investment would have allowed Alibaba to capture significant returns rather than allowing those returns to go to rivals. Instead, the AGI program asked for no financial commitments from the class members, no commercial relationship. In fact, at one point, two eFounders Fellows attending our program in Hangzhou approached me and my teammate, Dream Liu, and asked, "What is in all this for Alibaba?"

It was a reasonable question. They wondered why we would put time and resources into sharing this valuable digital knowledge without insisting on a measurable return on the investment. In truth, not everyone at Alibaba thought this was the best use of our resources and some questioned its purpose. There are, in fact, digital training programs run by other incubators or funds that insist on taking a percentage of equity in the participants' businesses in exchange for providing seed capital, advice, mentoring, and connections.

This was, in fact, part of the strength of the AGI program. Not only had these digital entrepreneurs flourished before joining the AGI program, they had also come up with ideas that were new to us. The "pupils" became "teachers" in many instances. Yet they all expanded the scope of the digital ecosystems, creating new opportunities, fueling inclusive economic development, and extending the reach of the digital economy. We were preparing them for a tomorrow that might or might not benefit Alibaba financially, but it benefited our company by furthering the success of our broader mission.

Our response to all those businesses that participated in AGI but brought no tangible payoff? As I commented to one eFounders class: "We hope you can become the Jack Mas of your own countries."

The New Newtonian Laws

One key to understanding the real-world impact and replicability of the Tao of Alibaba and the new Newtonian laws of the digital economy lies in comprehending that seeming paradox. What Alibaba might have "lost" by not turning the relationships with AGI participants into commercial transactions was more than countered by two related achievements that made Alibaba more effective and a better instrument for societal transformation.

The first was that we had fulfilled Alibaba's core mission in a vastly important way. We had widened the circles of opportunity by advancing those who had previously been sidelined in the global economy. We had accomplished one of Jack's key societal priorities, which he had articulated when he created Alibaba, empowering SME entrepreneurs by providing them the tools and market access needed to thrive. Our sense of purpose had paid off.

The second achievement was that these thriving ventures' real contributions to the digital economy in Africa was that their success became a catalyst. By demonstrating that their strategies worked, they helped attract more venture capital for themselves and others in the region. That inspired more homegrown entrepreneurs to build enterprises and gave African businesses, big and small, new markets to explore online. They accelerated the adoption of the digital empowerment model on that continent. They provided data that told entrepreneurs where to find new opportunities, what African consumers wanted, what support services would sell, and where bottlenecks or barriers existed and persuaded investors to supply the capital required to finance growth.

In short, the achievements of Africa's success cases expanded the pie for all the economic actors in that world, including Alibaba. More access to e-shops, a smoother flow of transactions, better logistics, more consumer spending power, and more positive government policy support were all products of this more inclusive and interconnected economic system. That was at the heart of what the AGI programs taught and why they were created.

The program implicitly conveyed the fundamental principle that the best companies are driven by a clear sense of social purpose, integrated into the company culture, and that the most durable gains are achieved when success is widely shared.

Training One Thousand Entrepreneurs

AGI—Alibaba Global Initiatives—was birthed, like many things at Alibaba, exceptionally fast out of one of Jack Ma's flashes of insight. Cai Song and I had been running AGLA, the internal global training program, teaching the company's new leaders about Alibaba's digital culture and learning quite a bit myself about the essence of our model and its impact. Things seemed to be going well. And then one day, in 2017, I was in Singapore on a business trip when my phone rang. It was Jack Ma's secretary.

"Brian, where are you at the moment?" she asked. "Jack wants you to join his meeting now with the secretary general of UNCTAD."

The secretary general of the United Nations Conference on Trade and Development was in Hangzhou, she explained, discussing models for inclusive economic development in emerging markets and how Alibaba and its foundation could help support UNCTAD's cause. In typical fashion, Jack suddenly had an idea on how we could advance those objectives, and he decided, on the spot, to put it into action. I was caught off guard because nobody had told me about the meeting.

"Ack, I'm in Singapore now," I told Jack's secretary. "But anything I can do to help?"

Suddenly, I was talking to Jack. "Brian, when you come back to Hangzhou, I want to talk to you," he said. "I want to create a program to help train a thousand young entrepreneurs from the emerging markets within five years. And I think you can help to do this. We will partner with the UNCTAD on this project. Bye."

That brief conversation began what eventually became the four AGI educational programs we created, with the flagship program, the eFounders Fellowship, operated in partnership with UNCTAD. Jack had been named UNCTAD's special adviser for young entrepreneurs and small business, which he took quite seriously, and now he was seeking to implement his vision.

That unexpected phone call and my new assignment became a pivot point, both for me and for the company. From Alibaba's earliest days, one of Jack's key growth strategies was education. He would travel around China and explain to small businessmen what the internet was, how it worked, and why they should put it to use by embracing e-commerce, even though the concept was little known and the online markets barely existed in China at the time. As a result, the new converts to his digital religion often utilized the Alibaba platforms, which fueled their explosive growth.

The new AGI program was different. Alibaba now made the huge conceptual leap from being an expanding digital success story, evangelizing to advance Alibaba's commercial interests, to being a facilitator in what is a global economic revolution of empowerment, inclusivity, and opportunity. In the end, some of these entrepreneurs may end up becoming Alibaba partners or even customers, but in reality, most will not. But such win-or-lose scoring has little relevance to this new iteration of Jack's vision and mission. And that, in the end, stands as a tribute to the paradigm's power.

The Roots of the Digital Ecosystem

Back in 2010, our vice chairman, Joe Tsai, responded to conversations within Alibaba about possible shifts in our business paradigm by circulating a letter to employees restating some of Jack Ma's principles. He reemphasized our embrace of the collaborative genius behind the digital economy. His statement remains a valuable explanation of the reality behind the new economy and informed our AGI programs.

Let me put this mission statement into perspective. What is a "new business paradigm"? In the "old" business world, businesses are isolated entities where they interact with suppliers and customers in a way that maximizes their own profit—managers are primarily concerned about how to squeeze their suppliers and make their customers pay more. It's a zero-sum game.

The Internet, which brought about speed and transparency of information, the ability to collect and analyze massive amounts of data, and the capability for people to collaborate real-time, has changed the game. We are living in a new commercial reality where the interests of participants in the value chain—from suppliers to producers to distributors to consumers—are more closely *dependent on each other* [my emphasis] than ever.

The Internet has enabled us to devise win-win solutions, and to thereby enlarge the size of the entire pie for everyone to share. In the new business paradigm, short-term profiteers, protected monopolies, nontransparent middlemen, unscrupulous producers and rip-off artists will eventually be driven out of the market. In the new business paradigm, those that thrive will be people who take a long-term view, share and collaborate within the ecosystem, and take responsibility for the well-being of customers, employees and the environment in which they are situated.

We did not "invent" the new business paradigm. It has emerged because the Internet necessitated its creation. In Chinese we use the term "新商业文明," which literally means a "new business civilization."

That is a clear statement of our culture that still holds true, though some have grown even more creative in articulating its essence. Catherine Mahugu, a Kenyan entrepreneur who created a digital platform for selling handmade jewelry and was a member of the first AGI eFounders Fellows class, put it this way: "There is an African saying that if you want to run fast, run alone. But if you want to run far, run together."

The promise of the Alibaba ecosystem model and its many global iterations is that we aim to run far, together.

Launching the AGI eFounders Fellowship Program

In November 2017, the first class of AGI's eFounders Fellows, put together under the joint sponsorship of the Alibaba Business School and UNCTAD, was drawn from seven African countries. It was a highly competitive process: there were thousands of applicants for the first twenty-four places. We spread the word and took nominations from, among others, business organizations, including venture capital firms and local business groups, such as chambers of commerce, trade associations, and government organizations. We cast our net widely, partly in recognition that we had much to learn about Africa and partly to make sure we identified the most dynamic and innovative entrepreneurs for our class.

The Fellows were already building companies focused on areas like e-commerce, digital payments, logistics, big data, and travel. We hosted them in Hangzhou for an intensive two weeks, where we provided a thorough explanation of the principles of the digital economy and how they could be deployed.

The curriculum covered the key characteristics of the digital economy in China, the preconditions required to build it, and the benefits it was creating in society. We engaged our Fellows in classroom presentations, discussions, and debates, through site visits, particularly to experience the Taobao Village phenomenon, and meetings with different units in the Alibaba ecosystem. The Fellows prepared final projects via a business hackathon that employed the tools and principles of the digital ecosystem.

We used Alibaba as the main case study for the group, and we walked the Fellows through the processes that led, one by one, to the creation of our chain of business units: Alibaba.com for connecting Chinese SMEs to the world's business buyers; Taobao for filling the gaps of the underdeveloped retail sector and spurring more domestic consumption; Alipay for addressing the financial infrastructure and trust gap in society and online business; Alibaba Cloud for providing low-cost, scalable computing power to a rapidly growing digital economy; and Cainiao for breaking the logistics bottlenecks in China, which were being exacerbated by increasing e-commerce activity.

We also visited innovation parks to show how new start-ups are emerging within the local ecosystems, companies that were providing services like livestreaming production or those focused on cross-border trade with markets in Southeast Asia and Africa. A highlight was a visit to Bainiu, with its edible nut–based digital ecosystem, to observe a Taobao Village in action and demonstrate the multiplier impact of e-commerce in a rural area. Nearly all the Fellows responded immediately and emotionally to this success because so many of them could envision villages in their home countries flourishing with access to a robust digital infrastructure.

Seeing firsthand how the e-commerce phenomenon had organically taken root was a powerful experience for the Fellows, which ended up stimulating numerous new entrepreneurial ideas, many beyond what we might have imagined. It was yet another example of how they took the digital tools we introduced and ran with them in directions we neither predicted nor expected. The pupils became the teachers with remarkable speed. This would be a running theme with each AGI group.

AGI Empowerment Model

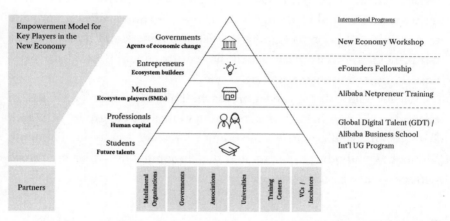

The layers and components of AGI's digital empowerment model.

Adapting the Concept

Of course, we understood that the Alibaba model could not just be transported and imposed on other developing countries without alteration: it would have to be adapted to the conditions in each market. To accommodate and support that reality, we focused on teaching the eFounders Fellows the fundamental principles behind Alibaba and our culture of collaboration and inclusiveness. We hoped this would be replicated in the fertile economic soil in Africa and elsewhere.

"We certainly do not expect you all to try and cut and paste everything from China and expect it to work in your home countries," I used to tell the AGI entrepreneurs. "Instead, it's more helpful that you understand the thinking and rationale behind our decisions along the Alibaba journey and what the key factors were that impacted the outcome. Most important is that you believe in your own abilities as an entrepreneur to impact change: that even in a market like China that started with nothing, Alibaba was able to create something significant by leveraging technology to change society."

Graduation projects were an attempt to get the Fellows to think about creative problem-solving. We ran a hackathon where they applied their newly acquired knowledge, using innovative partnerships and collaboration. We brought in judges, including senior partners from prominent venture capital firms and Alibaba business unit leaders, to provide feedback. In addition to helping the Fellows understand how to utilize our principles, this exercise helped the venture capital firms identify potential investment opportunities.

We gradually broadened our reach. Following that first class of eFounders Fellows from Africa, we brought in a second group, in March 2018, from Asian countries, including Indonesia, Malaysia, Thailand, Vietnam, Cambodia, Pakistan, and the Philippines, and then continued with two classes each year.

Expanding AGI

Successfully employing the ecosystem model requires cooperation among numerous entities, including government, entrepreneurs, traditional businesses, and educational institutions. (See AGI Empowerment Model on previous page.) This was an essential conceptual framework and one of the more significant takeaways derived from the AGI programs. Recognizing these communities of interest, we expanded our offerings and developed four programs under the AGI banner, all with a similar curriculum, to encourage the assembling of these digital building blocks.

The process often needs to start at the top with a commitment from government leaders to acknowledge the need for and benefits of investment in hard infrastructure, from roads and bridges to widely accessible internet connectivity, and innovation-friendly regulatory policies to support new ventures. Then the entrepreneurs must step in to analyze the marketplace and gather needed financial and human capital and build the platforms for trading products and services. These could include online retail markets, payment services, ride sharing apps, credit and banking services, and food delivery apps. Once the platforms are operating, consumers as well as traditional businesses generally need training or support to fully integrate the platforms into their everyday activities.

Finally, in order to power the development and overall growth of the ecosystems, there needs to be a sufficient amount of talent trained in the education system to support the growth. These four elements all played a critical role in building what is now the world's largest digital economy, China.

Thus, in addition to the eFounders Fellowship, we established what we called the New Economy Workshop to share with government officials from emerging economies an understanding of how the digital model evolved in China and what public policies can help the entrepreneurial sector flourish. We also started a program for what we dubbed Netpreneurs, traditional businesses with digital aspirations. The program also brought in digital enablers who are helping those traditional businesses make that transition to the online world. And we developed a fourth program in partnership with the Alibaba Business School to design an undergraduate e-commerce curriculum, which we provided to accredited educational institutions from Thailand to Kazakhstan and even Mexico.

The New Economy Workshop has played a significant role in the widening awareness and knowledge of the new paradigm. It has energized many government officials and given them a road map for how they can support inclusive development. This has also connected those officials with the expanding numbers of digital entrepreneurs in their countries and given them a common language for discussing useful policies and priorities. And, most importantly, it has helped them envision how the public and private sectors can collaborate to inspire continued digital transformation in their markets.

The Alibaba Netpreneur Training program included traditional businesspeople from a wide number of industries, including fashion and apparel, packaged foods, agriculture, furniture, industrial products, and even fisheries. The program also included digital enablers, companies whose mission is to assist the traditional businesses in their digital transformation. That combination provided an on-ramp, introducing those businesses to the platforms and ecosystems that the eFounders Fellows were operating in their countries.

To plan for the future, the AGI team also created the Global Digital Talent (GDT) program, which designed the curriculum for equipping college graduates with relevant skills. Those graduates provide the energy and ideas that keep the systems growing. Whether it is a Thai fish exporter utilizing e-commerce to sell more seafood abroad, or an Indonesian platform providing e-commerce tools for rural mom-and-pop shops connecting with urban consumers, these companies all need young talent to help them build. The students from these GDT programs are in high demand.

Building a Rwandan Digital Galaxy from a Textbook Lesson

Few have embraced the digital ecosystem model better or demonstrated the promise of the digital economy more than Clarisse Iribagiza, a remarkable young Rwandan entrepreneur and a member of a Netpreneurs class. Clarisse founded her company, HeHe, a supply chain platform and logistics service for SMEs, while still an undergraduate studying computer science at the University of Rwanda's College of Science and Technology. But her journey really began when, as a curious high school student, her economics teacher mentioned that, according to the classical textbooks, it was impossible to create a perfect market where consumers have complete information on all the goods and services available, how much they cost, and how to access them. Consumers are always at the mercy of market imperfections and incomplete knowledge of product choices and pricing, which inevitably translates into higher prices and a limited selection, she was told.

That thought stayed with Clarisse like a challenge, into college and then when she began a six-week technology business incubator program run by MIT. Bothered by these market imperfections, she decided this was a problem she wanted to solve, and the result was a digital business that cut through supply chain impediments and improved efficiency for consumers.

She was particularly concerned about inefficiencies in the global food system. As much as one-third of the fresh food in the supply chain spoils before reaching consumers, she found in her research, leading to wastage and destruction of value. So she designed a streamlined, data-driven supply chain system that cut down on those losses in Rwanda, improving farmer incomes and giving consumers more choice at lower prices. She went further in applying digital tools to build more predictability into the food supply system. HeHe created a "post harvest solution," which uses remote sensing technology to forecast food supply and consumer data to forecast demand. It can then efficiently allocate logistics resources, such as truck fleets and warehousing, to connect farmers to high value markets.

A key partner was the Rwandan government, which has built infrastructure and supported digital awareness among residents. Broadband internet connectivity is available in more than 90 percent of the country, and more than 70 percent of the Rwandan population have smartphones or access to them. To assist digital SMEs, the government provides entrepreneurs one year of free access to the internet, office space, and opportunities to learn from other digital entrepreneurs around the world.

But one of Clarisse's most powerful insights was that, leveraging her own success, she could become a partner to other young Rwandans eager to leap into the new digital frontier, especially women. First, she expanded HeHe's ecosystem by mapping out the supply chain for farmers and businesses and creating related businesses for logistics, warehousing, e-commerce, and digital payments to open new channels. That was only the start. She quickly found that one of the impediments she encountered was a lack of trained talent. As she related to me, she found in her country very few people who could translate ideas into functional applications that she could then scale. The solution? She opened an innovation academy to train potential coders, engineers, and staffers while they were still in high school. "This was one of the best decisions I ever made," she later told me.

She called her creation the HeHe Innovation Academy, and it was implemented in Rwanda's high schools and universities. The academy has trained more than 470 young people. Many contributed to the growth of her own company, but even more have gone on to found start-ups or lead technology organizations. Clarisse had the foresight to see that she helped her own company grow by ensuring that Rwanda and dozens of young, entrepreneurial Rwandans could thrive in the knowledge economy.

To further support entrepreneurs, Clarisse, together with other young ICT (information and communication technology) entrepreneurs, also launched the iHills, a network for Rwandan start-ups that provides mentoring, access to finance, and market information. The economic pie was expanding with each initiative.

When the pandemic struck, the wheels of innovation again went into motion. Clarisse saw opportunity in the enormous challenge of supply chain breakdowns to get food to needy families. She started Abundance Village, which tackled the disruptions by facilitating food distribution and ensuring a smoothly functioning agricultural system. AI, machine learning, and remote sensing tools were employed to anticipate where food was needed and match that with what farmers were growing. Overlaying the process was a newly designed service architecture so that the different digital platforms that already existed could easily speak to one another, therefore eliminating the need for all involved to learn to use numerous platform terms.

Asked once in an interview whether she saw herself as a dreamer or a doer, Clarisse did not hesitate. "I dream and I do," she said, adding a touch of ambition bordering on bravado and echoing the sentiment of Savio's yes theory in a Taoist sort of way. "Our mission is to create a perfect market to match demand and supply, therefore creating abundance for everyone in the world," she said. "Of course our current focus is Africa."

Clarisse's efforts have been matched by other Rwandan AGI program graduates. For example, a group of Netpreneur alumni teamed up and connected producers of high-quality coffee with Chinese consumers through Alibaba's Tmall Global shopping platform. Dean, the French

Alibaba colleague and AGLA graduate, assisted in building this program for Alibaba. In the first year, the sales of Rwandan coffee to China increased by 700 percent.

Origene Igiraneza, a Netpreneur graduate, worked with the Rwandan Ministry of Education and the Ministry of Innovation to utilize his online education platform, O'Genius Panda, to bring STEM education to more than 605 schools and 23,000 students in ways that had not been possible through the country's traditional education system, particularly during the COVID lockdown.

Finally, Dioscore Shikama, a member of the first eFounders Fellows class, has continued to develop his company, AgriGO, which assists Rwandan farmers in improving crop yields. It provides best practice insights on farming techniques and market data on crop pricing each season, all through personalized text messages in local languages. This ecosystem model has been picked up by numerous other graduates of the AGI programs, delivering replicable results in other countries.

Oswald Yeo, CEO and cofounder of Glints, an online job placement platform started in 2013 in Singapore, said the biggest impact of his experience as an eFounders Fellow was a mind-set shift. Previously, he said, he had focused on one product—helping place recruits with the right companies—but the Alibaba ecosystem model expanded his understanding of the role his enterprise could play both for job seekers and employers. Specifically, he started a skills educational platform to upgrade the quality of the talent he was assisting, creating a new layer of value.

"That enabled us to help more professionals, but at the same time it created a strategic supply-side advantage that drove more high-quality job matches within our ecosystem," Oswald said.

Glints now covers a number of Southeast Asian countries, and it experienced a rapid expansion with the turn to remote work during the pandemic. Glints can facilitate those remote arrangements by acting as the employer of record and handling payroll and compliance for certain remote workers. In short, he has opened up new types of opportunity to a growing workforce and is providing training and education, while helping employers expand the breadth of their talent pool.

The AGI program has been our attempt to disseminate the recipes behind the digital ecosystems and provide a springboard for their wide application. The eFounders Fellows have passionately embraced this sense of social purpose.

Digital Transformation in Malaysia

Malaysia has produced some of the best examples of this energetic embrace of the digital concepts. We achieved a breathtaking number of relationships across multiple levels through our AGI programs. Upon returning home from Hangzhou, the Malaysian eFounders set up their own business group, the New Economy Entrepreneur Association, and scheduled regular events, open to the public. The Fellows take turns presenting aspects of their learning from the AGI courses.

As an example, one of their programs, in August 2018, included the following presentations: "Building a Strong Ecosystem: Learnings from Alibaba," "Going Beyond Borders," "The Road to Digitalization: Business Transformation and Corporate Innovation," and, importantly, "Culture: Key to Business Transformation."

Some of our Malaysian graduates are now among the country's top entrepreneurs, running digital platforms for e-commerce, logistics, the gig economy, and fintech, including the firms EasyParcel (run by Clarence Leong), GoGet (run by Francesca Chia), and PolicyStreet (run by Wilson Beh). Several Fellows have been invited to join ministerial advisory councils, and three were brought in to brief the Malaysian prime minister. They shared what they learned about some of the common economic conditions in Malaysia and China and how the government could nurture Malaysian entrepreneurs.

By 2020, AGI had trained 24 Malaysian eFounders Fellows, who were all building digital platforms, and more than 130 Malaysian Netpreneurs, who represented traditional businesses engaged in digital transformations. We also certified five Malaysian universities in the e-commerce curriculum for undergraduate students, equipping thousands of fresh graduates with e-commerce knowledge and skills.

One night, I learned that a group of our Malaysian Netpreneurs had returned on a visit to China and were in Hangzhou. I met them at a nearby restaurant, a wonderful group that included K. C. Choi, who runs a company that does printing and packaging, and Ames Tan, whose electronics business refurbishes used computers. It was an animated, excited group, and they quickly got to the point.

"We have an idea, guys," Ames said to me and several AGI colleagues who had joined our meeting. "We want to follow through with our plan that we had presented at the hackathon competition during our course and truly replicate the Taobao Villages in Malaysia. We have assembled a team from our Netpreneurs class that can do this." This included Anna Teo, known as the "Musang Queen," whose company exports the Musang King variety of durian, a distinctive Asian fruit, to China, and Sean Lee, whose company does e-commerce marketing. Each had something to contribute to bringing e-commerce to neglected Malaysian rural villages.

The AGI team was impressed and eager to help. Working hand in hand, they coordinated resources across various Alibaba platforms and assisted in making introductions to collaborators to support the Netpreneurs in pursuit of their idea. In the following months, they refined their approach and partnered with the government to bring to life a framework they called DESA, which means "village" in Malay. It was to be a nonprofit organization providing digital training and assistance to underserved rural communities and help farmers and SMEs plug into domestic Malaysian e-commerce markets.

They identified the products that, they believed, would be most attractive to consumers and designed platforms to promote those products. Then they found a Malaysian aboriginal community in Bentong village that was capable of supplying the agricultural products in sufficient volume. They launched the effort by joining Alibaba's Lazada's platform as part of its famous Singles' Day promotion. All of the products in the promotion sold out by day's end.

News of the DESA program spread, and they were invited to promote DESA products in a special Tmall campaign called Malaysia Week, which was created specifically to introduce Malaysian products to Chinese consumers. I attended and spoke at a DESA event and was quite moved, witnessing how these entrepreneurs had translated their experience in the AGI program so quickly into an e-commerce initiative for traditionally marginalized Malaysian businesses.

News of other examples continued to reach us, where entrepreneurs seized on and expanded the digital ecosystem principles. One of the more inspiring was an on-demand gig worker start-up in Kuala Lumpur called GoGet. The company was founded by an eFounders Fellow, Francesca Chia, and two partners. What is impressive about the model they developed is that GoGet has constructed an entire web of support around gig workers rather than treating them as merely an inexpensive, flexible workforce for companies or individuals needing to accomplish simple tasks. The platform allows these workers to expand their gig assignment options, build on their collective experience, and develop a financial safety net to give them long-term security. It is one of the more pragmatic and empowering models we observed, transforming the individuals from what are often dispensable, largely faceless part-timers into individuals who are equipped to take control and shape their own futures. Moreover, by creating such a platform, GoGet is playing a role in digitizing the informal economy, bringing this entire working population into the mainstream and providing a safety net for a vulnerable population.

Francesca Chia grew up in Malaysia and England and received a degree in economics and Chinese studies from Northwestern University outside Chicago. She went on to work for a management consulting firm in Malaysia, but after a few years, she started to get restless for a new challenge, she said. She wanted to build a business that enabled people and solved a real social problem and landed on the idea of working with gig workers in the lower- and middle-income range, workers who often had a low level of skills and had difficulty in finding stable work situations.

The basic approach of the company is straightforward. It has three basic client baskets: multinational corporations in search of an on-demand workforce for new ventures, local corporations with temporary needs, and SMEs looking for everyday teammates to operate their businesses. The tasks range from deliveries and household chores to events work and administrative support. GoGet's innovation was to provide clients a more reliable and higher level of service than they might normally find in such part-time workers and to provide the workers—the company calls them GoGetters—the opportunity to build skill sets and find more sustained work opportunities.

The company does this by providing three levels of support to the GoGetters. All those who sign up with the company receive immediate training on how GoGet operates, the level of service required and how to provide it, and the company's culture. In addition, GoGet has a learning portal on its platform that allows workers to watch videos and receive training in specialized areas. They receive badges when they complete the courses, providing marketable job skills and valuable credentials when seeking new assignments.

Second, GoGet runs events and networking opportunities for the workers so they can expand their client bases, team up with other workers, and share useful information about work and practices. The company teaches them how to operate and expand their own self-employment businesses. All of this is provided free of charge.

An even more important distinction is that GoGet helps the workers build financial security. It provides training in financial literacy, again free of charge, helps workers obtain government insurance and pensions, helps GoGetters save directly through integrating the app with their savings accounts, and is presently working to create a solution to assist them in obtaining microloans if they need funding for their businesses.

Asked once in an interview about what advice she would give would-be entrepreneurs interested in her innovations, Francesca said the motivation should not be the financial rewards alone or the desire to chase the latest fad but to have a strong commitment to the people you are helping and the social consequences of the business.

"You want to always empower the people plugged into your ecosystem," she said.

The aim, she added, "is to genuinely care about the problem you are solving."

Spreading the Digital Gospel

"Fellow African entrepreneurs, take notice, the onus is on us to build a new Africa. No one will do it for us. Keep innovating, never grow weary, never get tired, never give up. Africa needs every hand on deck! No more conflict! No one person can achieve building this Africa. Everyone must do this!" The audience erupted with applause and cheers as Uju Uzo-Ojinnaka, an e-Founders Fellow, finished her inspirational speech, rallying the one thousand African entrepreneurs and attendees in the auditorium.

This was the first Africa's Business Heroes conference, an event co-organized by the Jack Ma Foundation and AGI in Accra, Ghana. The president of Ghana, the former UN secretary general Ban-Ki Moon, and Jack Ma were present to launch the conference and celebrate this new phase in the spread of digital initiatives. Together, we had planned this for the past year to introduce a competition for recognizing and encouraging Africa's most impactful entrepreneurs.

Uju began with her own story, how her futile attempts some years earlier to source groundnuts for an international buyer had motivated her to find a better way. She created Traders of Africa, a B2B marketplace as well as a financing platform to fund digital businesses in need of capital. Jack Ma's story had convinced her, she said, to take a chance and create a company to connect businesses online on a clear, functional platform, solving what was at that time one of Africa's biggest digital challenges. During her fellowship in Hangzhou she had had an "aha!" moment: she realized that partnerships were the best way to untangle the logistics, marketing, and financial impediments in the continent's marketplaces. Uju captivated the audience by recalling her journey and eventual success.

"We must open our minds to the true spirit of *ubuntu*, because together we can!" she concluded, using an African term meaning putting the interests of neighbors and community first.

A critical follow-up dimension of the AGI program was finding avenues for spreading digital ecosystem knowledge in each country and widening the circles of learning. Many Fellows, on their own, enthusiastically initiated communication efforts. They established digital economy learning groups to share their takeaways from the Alibaba training experience, not just with their internal teams but also with local business communities and government policy makers eager to facilitate this transformation.

Anum Kamran, founder of Buyon.pk, an online shopping marketplace for micro, small, and medium-sized businesses in Pakistan, has described how her sense of mission was renewed and redirected following her eFounders fellowship in 2018. The experience encouraged her to focus on empowering female entrepreneurs in Pakistan, she said. She has mentored women launching start-ups, organized workshops around Pakistan, and worked with trade groups and the government to resolve critical infrastructure issues, such as ensuring that online payments move swiftly from buyers to sellers and e-shops. She has become an advocate for the vitality of the digital space as the founder of the Pakistan E-Commerce Consortium, which is working to strengthen the e-commerce community and has encouraged the industry to take unified positions on infrastructure and regulatory issues. She also initiated another group called Activate, specifically to support women's issues in Pakistan. Her own business has sought to introduce mostly smaller, informal businesses to online markets in an effort to boost the sales and growth of these enterprises. Buyon.pk has registered more than five thousand sellers, 40 percent of which are informal businesses.

Anum attributes at least some of her success in building the digital community and ecosystem in Pakistan to a lesson she learned in her AGI program. The first thing she did on returning to Pakistan from China was to emphasize and share her company's vision with her employees, she said. Empowered with a clear sense of mission, the employees had a greater sense of ownership in their jobs and worked more efficiently, she said, and identified with the underlying cause. The end result was to free up more of her time so she could devote herself to her community building and mentorship projects.

How the COVID Crisis Strengthened the Digital Paradigm

When the COVID-19 pandemic swept the globe in early 2020, Clarence Leong's Malaysian package delivery and logistics business, EasyParcel, took a beating. Business ground to a halt. Clarence, a serial entrepreneur, had founded his business in Malaysia in 2014 and had guided his online venture to steady growth and learned about building a digital ecosystem as an eFounders Fellow. But it was all thrown into question by the pandemic lockdown, and EasyParcel experienced an initial 40 percent plunge in shipping volume during the first months of 2020.

Clarence received his education in the UK—paying his way, in part, by sourcing merchandise from China and selling it in the UK online—and had initially started his career as an aeronautics engineer. But he found working for a large aeronautics corporation unsatisfying, so he returned to Malaysia and launched online businesses. He eventually started EasyParcel to participate in the growing e-commerce marketplace. It thrived until it ran into the first pandemic disruption.

When e-commerce subsequently began to surge in the lockdown, and deliveries grew again, EasyParcel introduced a service called Pgeon Paperless to help customers suddenly forced to rely on the delivery service but who did not own printers. The new service allowed them to handwrite tracking numbers on the parcels they were sending, and Easy-Parcel devised a means of recording the data in digital form. In addition, the company started two other services to provide support during the economic disruptions. PgeonMart was an online grocery marketplace that provided a platform for small, local grocers to convert to e-commerce sales.

Also, to keep up with the growing demand for deliveries and to support the large number of unemployed workers, Clarence started Pgeon-FLEX. That was an online platform for quickly hiring the unemployed as gig delivery drivers. EasyParcel had, in effect, transformed into a platform for scores of what Clarence refers to as microentrepreneurs struggling to adapt in the lockdown.

As taught in the AGI eFounders program, Clarence provided a number of services to educate new customers and e-shop entrepreneurs in utilizing the platform. For new customers, his company provided online programs to teach everything from how to wrap parcels properly to how to book pickup and delivery. For the PgeonMart service, the company assisted shopkeepers in all aspects of establishing websites, including how to photograph, list, and price their products. The company also set up the EasyParcel x Pgeon Vaccination Campaign. Customers who could provide proof of vaccination received as many as twenty-five free EasyParcel shipments as an incentive to expand COVID-19 protection.

Clarence has also become an advocate for the digital community. He helped empower as many as ten thousand female online entrepreneurs—called the WomenPreneur campaign—by offering up to fifty free shipments to each participant. And when there were fears low-income groups couldn't afford meals due to job loss amid the COVID lockdown, his company arranged for the delivery of thirty thousand meals to individuals who could not access food supplies.

Another Malaysian e-commerce platform, StoreHub, founded by eFounders Fellow Wai Hong Fong, is a point-of-sale system for retailers that also responded to the pandemic with a series of innovations. StoreHub, in coordination with other eFounders Fellows, launched an online delivery service, Beep Delivery, to connect restaurants with their customers and maintain a flow of business. The service adopted StoreHub's software platform, giving each restaurant a user-friendly interface that relied on logistics companies GoGet, ZeptoExpress, and TheLorry, all founded by our Fellows. It was up and running within forty-eight hours of being initiated.

Wilson Beh, a graduate of our Netpreneurs Training program, has been an innovator in the insurance sector with his thriving business PolicyStreet. I had reconnected with Wilson during a get-together with a handful of other Malaysian Netpreneur attendees after delivering a speech at a local event in Kuala Lumpur.

We gathered in a restaurant following the conference, and Wilson explained how his AGI experience had encouraged him to focus on expanding access to insurance for the lower 40 percent income bracket in Malaysia, the so-called B40. Many workers at this income level could not afford standard insurance plans, he explained, yet they were the most vulnerable to catastrophes, such as costly illnesses or accidents. Wilson and his cofounders, Yen Ming Lee and Winnie Chua (also an Alibaba Netpreneur Training program graduate), had decided to devise a way to target that population on the PolicyStreet platform. The company now has a clear mission to make insurance protection simple and affordable with digital technology, particularly for the underserved.

One result of the pandemic lockdown was a steep increase in the use of delivery services, often relying on unemployed people transformed into gig workers. PolicyStreet quickly developed inexpensive insurance offerings for these delivery drivers. In addition, it created policies for SMEs whose owners and workers contracted the virus. It also established a COVID-19 insurance helpline and partnered with the Malaysia Digital Economy Corporation to provide savings on SME insurance renewals. They developed an interface on their platform that integrates thirty-five insurance companies whose policies are sold on the Policy-Street website, and they simplified the documentation process for everything from enrollment to filing claims. They have deployed data analysis and AI to provide real-time information on insurance trends, which has helped insurers design custom policies. And they obtained an underwriting license to develop their own microinsurance to provide protection for the digital and gig economy workers across Southeast Asia.

In the Philippines, Joshua Aragon and Steven Sy, eFounders Fellows from the sixth class, identified an opportunity to apply the AGI principles in their home country. Joshua had founded and was running Push-kart, an online grocery market, and Steven ran Great Deals E-commerce Corporation, which assisted companies in setting up e-shops. During the COVID-19 outbreak, we had a conversation with them about how they were adapting to the economic lockdown. They explained that they recognized how badly the crisis was impacting their communities, especially access to fresh food, and decided to act.

They launched a new venture, Zagana, an online platform that gives farmers, long subservient to an inefficient supply chain dominated by overbearing middlemen, direct access to Filipino consumers. Working with the department of agriculture, which sponsored visits to agricultural villages so they could observe conditions firsthand, they learned how poor market access, a lack of functioning infrastructure, and difficulty in obtaining financing for equipment or labor made life for the farmers harsh and the prospects for improvement slim. This was, among other things, driving young people away from their home villages due to the lack of opportunities.

Very quickly, they realized they needed a digital ecosystem to support their vision for assisting farmers and consumers. Orders soared from families unable to shop in person, but a broken logistics system meant numerous missed deliveries, order cancellations, and wasted produce. Joshua and Steven responded by developing their own network of small fulfillment centers for storing food, and they partnered with a digital logistics platform, Grab, to get deliveries directly to customers on time. Once the system was operational, the farm-fresh produce was often delivered within thirty minutes.

Not only has Zagana now expanded its partnerships with additional shopping and logistics companies to reach more consumers, it has created a sophisticated data analysis platform capable of assessing buying patterns to forecast demand, a boon for the farmers, and to pinpoint where they needed to open new fulfillment center hubs. They benefited from interest-free government loans for working capital. Within eighteen months, they had delivered more than one thousand tons of fresh food and helped five thousand farmers.

Just as the transformational character of Taobao Villages initially took us by surprise, the swift, highly effective responses of these digital entrepreneurs could never have been planned. It was a product of individual commitment to the public good and a sense of purpose, motivations that were given wing by the genius of the new economic paradigm.

That only underscored the truth of what we have learned, again and again, from the Tao of Alibaba and its application to the markets. *Talent, we know, is equally distributed among rich and poor, rural and urban dwellers, but opportunity is not.* The smartest thing that AGI did was not just in finding savvy entrepreneurs but also in spreading knowledge of this new economic model, which allows talented people to cultivate and apply their capabilities. The more the knowledge of this digital paradigm spreads, the more talented people will find it and seize the opportunities presented.

I was reflecting on the success of our AGI programs one evening as I sat in a car cruising through the rolling hills of Kigali, Rwanda, heading into town from the airport for some meetings. A light breeze played across my face as I took in scenery almost identical to the wonderful hilly landscape where my personal journey had begun, in the Bay Area of California. It felt so familiar.

The cool evening air brought a sense of tranquility, just as I remembered feeling so often on drives with my family as a kid. But now I was not in the heart of Silicon Valley, where everything digital seems possible. This was the heart of Africa, so long neglected and underserved, and I was on my way to participate in a gathering of a Global Digital Talent workshop that Alibaba had helped organize.

Through the AGI program, we had provided a framework and the principles of the digital ecosystem, but Rwanda was running so fast with the model that it was difficult to keep up. That was a particularly heartening, inspiring reality in a country so recently scarred by terrible violence but with a population now dominated by youth, ambition, and digital dreams. Through the AGI program we had engaged multiple players in the society, starting with the Rwanda Development Board, which had adopted a farsighted policy of taking practical steps to empower entrepreneurs, build out the digital infrastructure, and make Rwanda a digital hub for the region. Rwanda, in fact, has become one of the more attractive destinations for businesses trying to establish themselves in Africa, and educational institutions, such as Carnegie Mellon University and the Africa Leadership University, have set up campuses in Kigali, providing highly trained talent to bring this vision to life. The energy there is palpable.

As my car entered the bustling streets of the capital, I took some comfort in realizing that I was now an admiring spectator of this phenomenon. I took pleasure in hearing about the dreams of these passionate entrepreneurs, not so different from my own excited dreams when I had joined Jack Ma on those couches in his apartment nearly twenty years earlier, eager to learn about the new world taking shape before me.

12

The New Road Ahead

The Tao of 1s And Os

I have long admired how the visionary thinker R. Buckminster Fuller articulated a philosophical, almost spiritual faith in the power of technology to transform the human condition, to elevate human dignity and empowerment. He criticized some technologies—fossil fuels, for instance, and atomic energy—but few have believed more wholeheartedly, with an almost childlike exuberance, in the transcendent power of well-engineered, innovative tools to liberate people from the clutches of poverty, disease, and exploitation.

To be sure, his ideas, spun out in a nearly endless stream of lectures and books during his long life, can seem at times like fanciful intellectual meanderings. He once proposed putting the better part of Manhattan under a giant dome to cut energy use. In a thoughtful assessment, "R. Buckminster Fuller: Technocrat for the Counterculture," the Stanford professor Fred Turner called his notions "a cotton-candy of machine-age dreams." But few have expressed greater confidence in the belief that smart technologies could empower individuals, give human aspirations form and wing.

Pursuing his dream, Fuller created the geodesic dome and designed a new Dymaxion car and Dymaxion house. His thinking inspired some of the technical godfathers of the personal computer revolution as well as the builders of back-to-the-land communes in the 1960s, but he saw his job not as creating any one tool. His project was changing the arc of history, nothing less.

"We're going to start emancipating humanity from being muscle and reflex machines," Fuller wrote in his slender book *Operating Manual for Spaceship Earth*. "We're going to give everybody a chance to develop their most powerful mental and intuitive faculties." And that mission, in his view, was urgent. "We have discovered that we have the inherent capability and inferentially the responsibility of making humanity comprehensively and sustainably successful."

China's March from Rural Backwaters to Digital Ecosystems

Jack Ma and his transformative technological "tool," the Alibaba digital ecosystem, falls into that ambitious, visionary project. He had the good fortune, though, to have launched his business at the dawn of the digital age, and so Jack was not only able to embrace a philosophical dedication to individual empowerment and societal transformation but also to create platforms that have contributed to that kind of uplift on a historic scale. The Alibaba digital ecosystem benefited from and helped accelerate one of the greatest successes of economic advancement in history: China's meteoric rise over the past four decades from a deeply impoverished, struggling agricultural nation into what is now both an industrial and postindustrial powerhouse that is generating unprecedented wealth for its citizens.

When I first visited my Chinese ancestral village in 1985 with my father, as a middle school student from California, I had jetted from an affluent, modern society to a village in Taishan where commerce was powered by three-wheeled rickshaws and motorbikes, sputtering along under bulging gunny sacks and clucking chickens in wicker cages. I felt an emotional bond with the land of my ancestors but was startled by its humble state.

Fast-forward more than thirty years to the Taobao Villages Alibaba supported in once-impoverished rural backwaters and to the digital ecosystems that have now put China far ahead in its e-commerce growth. The rate of extreme poverty in China—once seen as an intractable reality—has plummeted. According to World Bank data, the extreme poverty rate in China—people living on the equivalent of $1.90 or less per day—declined from 88 percent of the population in 1981 to 36 percent in 1999 and 0.7 percent in 2015. Today, according to some studies, China has completely eliminated such extreme deprivation.

While the poverty eradication campaign has been a society-wide effort over many decades, it has been more recently fueled, in part, by finger taps on keyboards and smartphones, running on magical flows of electrons that would have seemed inconceivable in the subsistence economy of a couple generations ago. With mobile payment users in China reaching 872 million in the first half of 2021, mobile payments now constitute 50 percent of all payments, while cash now makes up only 13 percent, according to the China Internet Network Information Center. Meanwhile, eMarketer, a research firm, estimates that China proximity mobile payment users constitute close to 87 percent of the country's mobile phone users compared to 45 percent in the United States and 24 percent in the UK. In a 2019 article, CGAP (Consultative Group to Assist the Poor) noted that "digital payments are becoming so dominant that the People's Bank of China has had to forbid what it sees as discrimination against cash by merchants who accept only digital payments. This is all the more remarkable because just two decades ago, China was basically a cash economy."

China is becoming a leader in high value-added areas such as chips, smartphones, solar panels, and electric vehicle production, changing the contours of the global trading system. Better pay for workers, greater openings for Western imports, and shifting government policies have encouraged enormous increases in consumption, too, creating yet more entrepreneurial opportunities.

Perhaps the most significant dimension of the transformation is in how much of the wealth being created has gone to Chinese workers, small businessmen, farmers, and women. That was a core element in Jack Ma's mission, more inclusive prosperity. Alibaba represents a case study of how one company has managed to create a positive social impact that most would have said was impossible. How could a company in a country with so little to begin with build a digital ecosystem of such abundance? That is a powerful question I have tried to address.

But the benefits have not gone just to the Chinese. Alibaba teams all over the world have worked to help make that transformative platform of 1s and 0s one of China's more valuable exports, whether through education or our local business platforms in emerging markets in Southeast Asia, South Asia, the Middle East, Africa, and elsewhere. The AGI program has trained hundreds of entrepreneurs in how to create and deploy digital ecosystems, replicating the success we achieved in China. Even better, our work at AGI merely planted the seeds of innovation. The entrepreneurs themselves not only replicated the success of the digital economy but also enlarged on what we taught them. The students have been educating their teachers.

But the local entrepreneurs there are also employing digital sorcery to generate more inclusive growth in countries where the old economic models had failed to generate as much of a lift. Replicability of this digital model has become evident in the rise of a plethora of new regional platforms from India to Indonesia. Players such as Tokopedia and Shopee have filled the need for a retail commerce platform not unlike Taobao in China, innovative payment platforms like Paytm, GrabPay, and DANA digital wallet have filled a similar role as Alipay, while GoTo (formerly Gojek), EasyParcel, and Shipper have created their own logistics network suited for local market characteristics. These and many of the AGI entrepreneurs highlighted in previous chapters are just a few examples reflecting what Fuller meant when he talked about using technology for "emancipating humanity" and are the fulfillment of a major element in Jack Ma's founding vision for Alibaba.

The Yin and Yang of Digital Opportunity

The Tao of Alibaba is a phrase for numerous characteristics of the Alibaba model: how it integrates a host of elements to align with a company's mission and vision, how it aims to follow the Taoist principle of operating in harmony with the environment's natural course or the way, and how Alibaba's principles are often a marriage of opposites, a sort of strategic dialectic that drives constant change, constant re-invention, a unity of contradictions, yin and yang in opposition and partnership.

What are those opposites? Alibaba is a very large organization that values the small, individual entrepreneur. Its sister company, Ant Group, one of the largest fintech companies, derived its name from the idea that, individually, a single ant can do little, but an ant colony, a collective, is capable of building or moving mountains. Alibaba is in a constant state of becoming, adapting, and changing, but it has stayed faithful to a steady set of principles.

And Alibaba is a for-profit enterprise that naturally seeks to boost returns, but it does so in the service of what it perceives to be the common good, broadly shared societal improvement—in other words, as Jack often says, creating an organization guided by a charitable heart but driven by business mechanisms.

The Taoist philosophy emphasizes interconnections, and Jack was an architect of one of the great feats of interconnectedness by putting so many people and small businesses on the web for the first time. Discerning the way means resisting separateness and division and appreciating the natural connections of the world. As Puett and Gross-Loh wrote, "The more we see the world as differentiated, the more removed we become from the Way. The more we see the world as interconnected, the closer we come to the Way."

Tens of millions of businesses, many previously too small to participate in global trade, and over a billion consumers, lots of them previously unbanked and disconnected from the mainstream economy, have now been linked thanks to Alibaba and other digital platforms. Alibaba thrived on these links and further strengthened itself by teaching this model to entrepreneurs in numerous emerging markets.

This ecosystem model rejects the zero-sum philosophy that used to be prevalent in the business world. The new model has proven, over and over, that collaboration, sharing, transparency, mentoring, and mutual support are more successful approaches for accelerating development and inclusive wealth creation in the digital economy. This, too, is very much in line with one of Bucky Fuller's farsighted appeals.

"So, planners, architects, and engineers, take the initiative," Fuller wrote. "Go to work, and above all co-operate and don't hold back on one another or try to gain at the expense of another. Any success in such lopsidedness will be increasingly short-lived. These are the synergetic rules that evolution is employing and trying to make clear to us. They are not man-made laws. They are infinitely accommodative laws of the intellectual integrity governing the universe."

The Uncarved Block

Taoism, of course, cannot be reduced to just these few principles. It is a complex system, a thought process, a sometimes enigmatic assemblage of ideas that have survived for centuries because they have offered a guide of sorts for living life with meaning. Laozi, regarded as the great philosopher of Taoism, whether he was a real historical character or not, often resorted to riddles to express the dialectic behind the principles.

"If all people of the world know that beauty is beauty, then there is ugliness," he wrote. "If all people of the world know that good is good, then there is already evil."

One of his more compelling expressions was an idea written in Chinese as *pu* (朴), which can be defined as "natural, simple" or "honest." As a Taoist concept, it is sometimes translated as "the uncarved block." In other words, it suggests, among other things, something with innate potential. The potential is in the nature of the block. We can regard this as the possibility, promise, and opportunity in our lives.

As mentioned in Chapter 8, at Alibaba, we sometimes described this critical trait as a simple and innocent quality, which essentially meant being positive, open to change and fresh approaches. What I learned after almost twenty years at Alibaba is that this sense of youthful confidence is part of our Tao. It suggests the optimism behind our platforms. We often spoke inside the company of how much we cherished this innocent and naïve culture as a source of our confidence that anything is possible.

Most of the entrepreneurs attending the AGI training programs arrived expecting just to hear about Alibaba's latest business models and the technological applications of commerce, payment, logistics, and big data. But they were surprised instead to learn in the process how the Tao of Alibaba placed more emphasis on ethos—the interactions among the softer, more subtle aspects of a company's purpose, objectives, strategy, organization, and people and how they must all interact and strike a balance.

The reality of understanding the tao is that, much like Jack's own life journey, it requires trial and error, real experience, and then conscious iteration before you can master the process. At the same time it requires persistence and resilience to endure such a journey. That persistence is highly reliant on whether the entrepreneur and his or her team have a strong sense of mission and vision. Numerous entrepreneurs told me that this simple realization helped liberate their thinking and unleash a whole new mind-set filled with inspiration. And this was even after some of them had been operating their companies for years.

There will inevitably be challenging times along the journey, times when you feel overwhelmed by factors out of your control. It is, therefore, critical that we all take the time as founders or leaders of our organizations to periodically ask the important question of why our company exists and whether we have aligned all of the critical components of the tao to fit that purpose. Once that question is answered, you and your team will be well anchored, you will be more able to resolve the seeming contradictions before you, you will be able to shape the uncarved block, and, ultimately, you will enable your enterprise to grow and to evolve in a more cohesive and coordinated way.

In hindsight, what our success in teaching the digital model through the AGI program demonstrated, repeatedly, is not just that it is replicable but that it is actually evolving beyond much of what Alibaba created. The model promises newer and more interesting iterations each time a digital entrepreneur seeks to solve a problem and invents a creative new approach in line with the conditions in that country.

Change is the unchanging reality of the digital world, and a series of new developments already promise a fundamental shift in how digital commerce operates, opening significant new opportunities for even more rapid, inclusive economic growth. This new internet iteration is often called Web 3.0, or web3. Among the early and most visible manifestations of this new approach are crypto currencies and the blockchain technology behind those currencies, what many refer to as decentralized finance (DeFi). Indeed, fintech (or internet-powered finance) stands in the spotlight today because of the fundamental position it plays in transforming so many aspects of the international economic system and also everyday life. But the innovative Web 3.0 decentralized architecture, in sharp opposition to the centralized structure of the current system, Web 2.0, will open more doors of opportunity to entrepreneurs and visionaries.

Web 3.0 Reset

It is too early to determine how Web 3.0 will perform in practice. Web 2.0 certainly delivered benefits in terms of connectivity and information democratization, yet it also brought about the emergence of problems some refer to as "surveillance capitalism" and potential threats to privacy, among other issues. All technologies pose threats and risks, often unanticipated. These concerns have been underscored by discussion around the enormous concentration of power and money in the hands of a small number of social media, search, and e-commerce giants.

There are debates over issues such as violations of privacy, the harmful spread of disinformation, hacking, identity theft, and ransomware attacks and monopolies or near monopolies that suppress competition and innovation. We are currently witnessing a backlash, as governments in the United States, the EU, and China investigate suspected abuses and explore and implement strategies for reducing the dominance of these companies, from imposing new regulations and major fines to possibly breaking them up.

Even in this early, experimental phase, there is a growing consensus on the prospects of Web 3.0 for establishing new directions in the digital economy.

Bernard Marr, a futurist and author of the book *Big Data in Practice: How 45 Successful Companies Used Big Data Analytics to Deliver Extraordinary Results*, recently explained that Web 3.0 will lead to "data democratization," the idea that "everybody has access to data and there are no gatekeepers that create a bottleneck at the gateway to the data." Ultimately, Web 3.0 will impact how industries like finance, commerce, technology, and the media will evolve, displacing intermediaries and enabling producers and creators to capture more value and gain more control over their destinies.

Pitchbook, a financial data company, reported that venture capitalists have invested more than $27 billion in crypto-related start-ups globally, a sum totaling more than the amount invested in the same sector in the previous ten years combined. While some are undoubtedly investing because of the prospect of enormous returns, socially conscious entrepreneurs are focusing on the hope that data democratization will benefit SMEs, consumers, and the overall economy.

In a report from October 2021, "How to Win the Future: An Agenda for the Third Generation of the Internet," the venture capital firm Andreessen Horowitz characterizes the new model as something of a curative for the abuses and failings of Web 2.0.

The Web 3.0 technologies can potentially address those problems because they are decentralized and transparent, owned by their users, and support more innovation. They could potentially reduce ransomware attacks, for instance, and accelerate transparency in the financial system by providing access to greater volumes of financial data much faster. Even creativity in the arts may flourish as never before, according to the report.

With Web 3.0, misaligned incentives between platforms and creators could be addressed by broadening access and dissolving silos between creators and fans. Through NFTs (non-fungible tokens), digital payments, and tokens, for instance, musicians and content creators have the ability to transact over the internet, directly with their fans, capturing more of the value of their work.

"Certain platforms are also experimenting with fractionalized ownership, allowing a community to pool its resources and collectively own a multimillion dollar artwork, with individuals then able [to] buy and sell their fractional stake to other individuals without the community selling the underlying artwork," the report said. Perhaps Web 3.0 will bring about a renewal of creativity and innovation, the report concludes.

That is high-minded—and speculative. In the real world, such glimmering hopes rarely arrive untarnished by baser realities. But there is no doubt about the billions of dollars being invested in Web 3.0 ventures because of their promise and the remarkable opportunities they are offering entrepreneurs, including the kind Alibaba has worked with in Africa and Asia.

Just one indication of the new sense of freedom is the new type of corporate organization it is supporting. In contrast to the traditional, hierarchical organizational model, these new entities are referred to as DAOs, or decentralized autonomous organizations. As the name suggests, they are built around rules on transparent, shared software stored on a public, permissionless blockchain without central control and without strict boundaries. They are collectively operated.

"DAOs enable individuals to collaborate, manage projects, own assets, invest, and operate like a traditional organization, but they can provide far greater levels of transparency, openness, and democratic governance," according to Andreessen Horowitz.

Only time will tell whether the reality meets the expectations. And whether or not Web 3.0 will become the corrective for all the challenges associated with Web 2.0 is a big question. Part of the problem is that the key will be how Web 3.0 solves tomorrow's problems, not just today's. As Jack has alluded to on multiple occasions, what's critical is that the technology we develop today is not just designed for the present but also for the future, keeping in mind market and societal needs now and moving forward.

China and the United States, Yin and Yang

Venture capitalists, entrepreneurs, Alibaba, and would-be e-commerce merchants confront these questions with the geopolitical environment growing more uncertain and contentious. The biggest digital enterprises are both a cause of some of the uncertainties and are victims of it. There is also political jousting within and among countries over the directions our digital future will take. Hence, the efforts to decouple the two economies will have repercussions for the entire world.

This is the backdrop to the current competition between China and the United States. Perhaps in a reflection of Taoist principles, the two countries are locked in an uneasy dialectical embrace. They are deeply intertwined—intensive collaborators and commercial partners across a broad array of industries—with the academic, scientific, and business alliances between two countries greatly benefiting not just the two nations but the entire world. Yet at the same time, the relationship of these two countries has also become one of archcompetitors—engaged in struggles over everything from regional military power to human rights, cyber spying, intellectual property, and spheres of global influence, thereby creating an atmosphere of distrust, animosity, and confrontation.

As the beneficiary of education and life experience from the United States and China, I have a great appreciation for what both have given me, their special qualities and how I have been able to absorb values from each. I worry about this competition and acrimony and the prospect of them causing economic disruptions and more importantly depriving the world of the overall benefits that have resulted from the constructive collaboration between the two nations and their peoples. Perhaps this is yet another situation where we can apply the principles of the tao to see that the answer is neither black nor white but a blend of the two, which will lead to the better outcome. In other words, we need to find a productive path between opposing perspectives.

Jack took inspiration from the spirit of Silicon Valley while applying his own methods to addressing the challenges he initially found in China. Today, the United States and more of our local American entrepreneurs might take inspiration from China's digital transformation, particularly in prioritizing inclusive development and spreading more of the benefits of the tech boom to underrepresented communities domestically but also abroad.

Final Thoughts

What is critical is that this next iteration of the digital economy— exciting, unsettling, and ponderous as it may prove—does not change the fact that it's just a tool. It provides a means of achieving objectives, but what matters most is the values that lie behind those objectives and how useful and inclusive they are. Perhaps this new Web 3.0 paradigm could accelerate the empowerment strategies that Jack Ma launched at Alibaba or, at the least, create the groundwork for what might be called a digital economy reset.

Jack may have articulated the idea differently, but when he counseled me after my entrepreneurial detour in Hong Kong and urged that I travel into China's poorest region, actively seeking discomfort, I believe he was urging a reset of sorts. He wisely saw that I would benefit from a journey that forced me to contemplate my values—not metrics or business principles—so that I would regain a clear understanding of the impact I wanted to have. Being "uncomfortable" was a chance to rediscover that motivation.

Let's hope this period of discomfort we are undergoing in today's world is a precursor of a reset on a larger scale. But in the meantime, it might help to start with ourselves as we look for answers in a more Taoist way.

As Jack often says: "If you want to change the world, you can start by changing yourself."

Author's Note

This book is the culmination of nearly two decades of experience working inside a company, observing, sometimes contributing, but more than anything reflecting on how this phenomenon called Alibaba unfolded. These reflections are my own opinions and interpretations of my experiences while working at the company, and the book was written independent of Alibaba Group.

For full disclosure, I remain a shareholder of Alibaba Group, and I am an angel investor in two start-ups referenced in this book: Glints and PolicyStreet. I am pleased to be supporting companies that pursue social missions with such commitment and dedication, and I hope this book persuades others to support entrepreneurs and organizations that are doing the same.

With regard to the romanization system used for spelling Chinese words in this book, I have mainly applied the Pinyin system. I have made exceptions in reference to certain book titles and specifically the word *tao*, since it remains a more commonly used form in English.

Acknowledgments

First and foremost, I would like to thank my wife, Chaai, who has been my unwavering bedrock of stability and a good counselor, not just for the content of my book but also for managing and prioritizing my life tasks amid a tumultuous time of COVID, launching a start-up, and making a major life transition with two new children. Without her unconditional love and support, none of this would have been possible.

Next, I would like to express my appreciation to my literary agent, Leah Spiro, who, despite the unforeseen change of events following the securing of my first book deal, continued to believe in the value of this evolving project over the next ten years and provided essential advice to help me navigate this journey all the way to today. I am also grateful to my publishing editor, Colleen Lawrie at PublicAffairs, for giving me this opportunity to share something I hope will be of great value to entrepreneurs and those who want to improve the world in the United States, China, and beyond.

A heartfelt thank you to my Alibaba Global Initiatives team: Cai Song, Dream Liu, Chandee Zhuang, Lusha Chen, Dan Liu, Loren Newman, Gareth Wetherill, Anna Yan, Matthew Gong, Cathy Hu, Jason Xiao, Rachel Wu, and others. All these colleagues put their heart and soul into building our AGI programs and played a key role in designing and executing the coursework, which resulted in some unforgettable years together and provided much inspiration for this book.

I am extremely grateful to Jim Sterngold, who played an invaluable role as a collaborator and adviser through the most challenging times of this writing process. His sage counsel and brilliant edits aided greatly in the creation of this work. Thank you also to Daniel Huang and Noelle Mater for their tireless efforts and contributions to this process, to Sicheng Zhong for his research assistance, to Anne Greenberg for her copyediting support, and to Gareth Wetherill for his wonderful graphic designs.

A special acknowledgment to Professor Ming Zeng, who, as chief strategy officer and dean of the Alibaba Business School, encouraged me to pursue this project because he recognized the value it would have for those inside and outside Alibaba. Jane Jiang, one of Alibaba's original founders, provided useful insights when recalling Alibaba's past events and offered moral support through the writing process. Savio Kwan, my revered mentor from Alibaba's early days, has always been there to provide guidance throughout my career at Alibaba. I appreciate the sessions we spent together in Hong Kong to discuss topics of leadership and mission, vision, and values. Alibaba colleagues Justine Chao, Yan Meng, and Roger Zhang and Ant colleagues Jason Pau and Fanny Wu were all very supportive. Finally, Porter Erisman, my "partner in crime" during the early years at Alibaba, provided insightful comments on my manuscript and he, along with former colleagues Annie Xu and Abir Oreibi, helped refresh my memory of the company's history along the way.

I am grateful to all of the eFounders and AGI entrepreneurs who took the time to share with me their compelling and inspiring stories, some of which were recollected in the book, but so many more that were not able to be included but also deserve recognition. Thank you also to Arlette Verploegh Chabot and UNCTAD for their support in helping to hatch our first AGI training program, the eFounders Fellowship in 2017, which provided the momentum and impetus to expand into the full AGI programs. Appreciation also goes out to Tom Tsao of Gobi Ventures, Sam Gichuru of Nailabs, Khailee Ng of 500Global, and many others for their support of AGI's programs.

I would be remiss if I did not acknowledge my father, who took great interest in my book project and provided a constant source of useful feedback on issues big and small. His deep wisdom and unwavering support provided an important foundation for the long and sometimes arduous journey.

Finally, the two individuals to whom I cannot express enough gratitude and appreciation are Joseph Tsai and Jack Ma.

Joe offered me the opportunity to be part of an adventure so unique and meaningful that it changed the trajectory of my life. He has been one of my greatest role models, someone who is deliberate in his thoughts and actions, upholds the highest values in all that he does, and remains one of the most grounded and humble individuals I know.

Jack has provided the light for what is possible to achieve in one's life no matter background or circumstance. He also generously provided a window into his world where I observed through his interactions with people from all walks of life that one of his greatest strengths as a leader is empathy and compassion. In short, my time with Jack cemented the belief that approaching life in pursuit of ideals is not futile. There are ways to do good for society while also serving the needs of a business, and with the right mind-set, dreams actually can come true.

Bibliography

Chapter 2

Cable.co.uk. "Worldwide Mobile Data Pricing in 2021." 2022. www
.cable.co.uk/mobiles/worldwide-data-pricing/.

China Daily. "10 Zhejiang Socioeconomic Achievements of 2021."
China Daily, March 4, 2022. http://ningbo.chinadaily.com.cn/2022-03/04
/c_722114.htm

Clark, Douglas. "China Becomes First Country in Which Ecommerce
Surpasses 50% of Retail Sales." Insider Intelligence, February 17, 2021.
www.emarketer.com/newsroom/index.php/china-becomes-first-country
-in-which-ecommerce-surpasses-50-of-retail-sales/.

Clark, Duncan. *Alibaba: The House That Jack Ma Built*. New York:
HarperCollins, 2016.

Congressional Research Service. "China's Economic Rise: History,
Trends, Challenges, and Implications for the United States." Updated June
25, 2019. www.everycrsreport.com/reports/RL33534.html.

Dongye Qimeng, Janus. "If China Has so Much Money to Invest in Other
Countries, Why Don't They Develop the Poor Parts of China?" Quora,
May 7, 2019. www.quora.com/If-China-has-so-much-money-to-invest
-in-other-countries-why-dont-they-develop-the-poor-parts-of-China/answer
/Janus-Dongye-Qimeng.

Economist staff. "The Great Mall of China: The Next Big Thing
in Retail Comes with Chinese Characteristics." *Economist*, January 2,
2021.

Global Times staff. "China Has 1.032 Billion Internet Users, 73.0%
Penetration Rate." *Global Times*, February 25, 2022. www.globaltimes
.cn/page/202202/1253226.shtml.

Higgins, Tucker. "China Surpasses U.S. as Largest Recipient of Foreign Direct Investment During Covid Pandemic." CNBC, January 24, 2021. www.cnbc.com/2021/01/24/china-received-more-foreign-investment-last-year-than-us-un-says.html.

Lardy, Nicholas. "The Changing Role of the Private Sector in China." Reserve Bank of Australia Conference paper, 2016.

Lugo, Maria Ana, Martin Raiser, and Ruslan Yemtsov. "What's Next for Poverty Reduction Policies in China?" Brookings Institution, September 24, 2021. www.brookings.edu/blog/future-development/2021/09/24/whats-next-for-poverty-reduction-policies-in-china/.

McCarthy, Niall. "The Countries with the Most STEM Graduates." *IndustryWeek*, February 6, 2017. www.industryweek.com/talent/article/21998889/the-countries-with-the-most-stem-graduates.

New China TV. "Jack Ma and Elon Musk Hold Debate in Shanghai." New China TV, August 29, 2019. YouTube video, 46:52. www.youtube.com/watch?v=f3lUEnMaiAU.

People's Daily Online. "China Had 8 Million Internet Users at the End of 1999." *People's Daily*, January 12, 2000. http://en.people.cn/english/200001/12/eng20000112T122.html.

Schwab, Klaus. *The Fourth Industrial Revolution*. Geneva, Switzerland: World Economic Forum, 2016.

Shaohua, Jia. In-person conversation with author and class visit in Yiwu, May 2017.

Shi-Kupfer, Kristin, and Mareike Ohlberg. "China's Digital Rise: Challenges for Europe." MERICS, April 8, 2019. https://merics.org/en/report/chinas-digital-rise.

State Council, PRC. "State Council Releases Five-Year Plan on Informatization." People's Republic of China, December 27, 2016. http://english.www.gov.cn/policies/latest_releases/2016/12/27/content_281475526646686.htm.

Statistica. "Penetration Rate of Mobile Internet Users in China from 2008 to 2021." Graph, March 14, 2022. www.statista.com/statistics/255552/penetration-rate-of-mobile-internet-users-in-china/.

Statistica. "Retail E-commerce Sales in China for 2021 with Forecasts Until 2025." Graph, March 22, 2022. www.statista.com/statistics/289734/china-retail-ecommerce-sales/.

United States Postal Service. "Market Dominant Products Final Revenue, Pieces, and Weight by Classes of Mail and Special Services." Fiscal year 2020. https://about.usps.com/what/financials/revenue-pieces-weight-reports/fy2020.pdf.

WIPO. "China Tops Patent, Trademark, Design Filings in 2016." World Intellectual Property Organization press release, December 6, 2017. www .wipo.int/pressroom/en/articles/2017/article_0013.html.

Wong, Alvin Cheuk Him. "A Study on the Credit Card Market in China: Influence of Service Provision Point and Credit Card Acceptance Point." Bachelor honors thesis, Hong Kong Baptist University, 2005.

World Bank. "GDP per Capita (Current US$) China." Graph, October 10, 2021. https://data.worldbank.org/indicator/NY.GDP.PCAP.CD?end=2020 &locations=CN&start=1999.

World Bank. "School Enrollment, Tertiary (% Gross)—China." Graph, September 2021. https://data.worldbank.org/indicator/SE.TER.ENRR?loca tions=CN.

Yang, Hongliang. "Overview of the Chinese Electricity Industry and Its Current Issues." Working paper, University of Cambridge, February 2006. www.repository.cam.ac.uk/bitstream/handle/1810/131663/eprg0517.pdf %3Bjsessionid%3D5B82CA7CFC218475FC30ECDD1E5AE4C9?sequence %3D1.

Yang, Yingzhi. "China Surpasses North America in Attracting Venture Capital Funding for First Time as Investors Chase 1.4 Billion Consumers." *South China Morning Post,* July 5, 2018. www.scmp.com/tech/article/2153798 /china-surpasses-north-america-attracting-venture-capital-funding-first-time.

Chapter 3

Alibaba Cloud. "A Brief History of Alibaba Cloud Apsara System." Blog, July 23, 2018. www.alibabacloud.com/blog/a-brief-history-of-alibaba -cloud-apsara-system_593843.

Alibaba Group. "Dream Maker." Company video, August 23, 2019. www.bilibili.com/video/BV1J4411R7tE/.

Alibaba Group. "Global Shopping Festival 2020." Fact sheet, November 2020. https://osssource.alizila.com/uploads/2020/11/2020-11.11-Global -Shopping-Festival-Master-Factsheet-Final-1.pdf.

Business Wire. "Ant Financial Unveils China's First Credit-Scoring System Using Online Data." Ant Financial press release, January 27, 2015. www.businesswire.com/news/home/20150127006582/en/Ant-Financial -Unveils-China%E2%80%99s-First-Credit-Scoring-System-Using-Online -Data.

Carsten, Paul. "Alibaba's Singles' Day Sales Surge 60 Percent to $14.3 billion." Reuters, November 12, 2015. www.reuters.com/article/us-alibaba -singles-day/alibabas-singles-day-sales-surge-60-percent-to-14-3-billion -idUSKCN0SZ34J20151112.

Custer, C. "Jack Ma on the Origins of Alipay and Learning to Say No." Tech in Asia, May 26, 2015. www.techinasia.com/jack-ma-origins-alipay-learning.

ECOVIS R&G Consulting, Beijing. *E-commerce in China*. Industry report, 2014. https://web.archive.org/web/20171018160231/https://www.pfalz.ihk24.de/blob/luihk24/international/Greater_China/China/downloads/2755990/f73ca66a4452cb651229217a2c72265b/E-Commerce-in-China-Broschuere-data.pdf.

Federal Reserve. "Credit Cards: Use and Consumer Attitudes, 1970–2000." *Federal Reserve Bulletin,* September 2000. www.federalreserve.gov/pubs/Bulletin/2000/0900lead.pdf.

Flora, Liz. "How US Tech Giants Copy China." Daily Insights, Gartner, May 31, 2019. www.gartner.com/en/marketing/insights/daily-insights/how-us-tech-giants-copy-china.

Gelles, David, and David Yaffe-Bellany. "Shareholder Value Is No Longer Everything, Top C.E.O.s Say." *New York Times,* August 19, 2019. www.nytimes.com/2019/08/19/business/business-roundtable-ceos-corporations.html.

Kang, Tae Soo, and Guonan Ma. *Growing Credit Card Markets in Asia: Challenges to Policymakers*. BIS Research and Publications, April 3, 2009. www.bis.org/repofficepubl/arpresearch_fs_200806.10.pdf.

KPMG. *On the Move in China: The Role of Transport and Logistics in a Changing Economy*. Report, 2011. https://assets.kpmg/content/dam/kpmg/pdf/2011/12/Transport-Logistics-in-China-201112.pdf.

Lunden, Ingrid. "Alibaba Dominates Mobile Commerce in China, with 76% of All Sales and 136M MAUs." TechCrunch, May 7, 2014. https://techcrunch.com/2014/05/06/alibaba-dominates-mobile-commerce-in-china-with-76-of-all-sales-and-136m-maus/?guccounter=1.

Ma, Jack. Talk during Davos forum, January 2015.

Millward, Steven. "Just Short of 2 Years Old, WeChat App Surpasses 300 Million Users." Tech in Asia, January 16, 2013. www.techinasia.com/confirmed-wechat-surpasses-300-million-users.

Newton, Casey. "Mark Zuckerberg Says Facebook Will Shift to Emphasize Encrypted Ephemeral Message." The Verge, March 6, 2019. www.theverge.com/2019/3/6/18253458/mark-zuckerberg-facebook-privacy-encrypted-messaging-whatsapp-messenger-instagram.

Routley, Nick. "Meet China's 113 Cities with More than One Million People." Visual Capitalist, February 6, 2020. www.visualcapitalist.com/chinas-113-cities-one-million-people-population/.

Stat, Nick, and Shannon Liao. "Facebook Wants to Be WeChat." The Verge, March 8, 2019. www.theverge.com/2019/3/8/18256226/facebook-wechat-messaging-zuckerberg-strategy.

Thomas, Denny, and Paul Carsten. "What's Up for Alibaba's Mobile App Strategy?" Reuters, February 25, 2014. www.reuters.com/article/us -mobile-alibaba-idUSBREA1N1S020140224.

Tsai, Joseph. Alibaba international staff communication re: Alibaba Group Mission, April 23, 2010.

Wolf, Titan. "Wang Jian: Why the Secret of Alibaba's Double 11 'Go to IOE?'" Karatos. https://titanwolf.org/Network/Articles/Article?AID =81563ec1-c541-4676-913e-2e558e3a0231#gsc.tab=0.

Zhuang, Chandee. AGI eFounders Fellowship lecture on Alibaba ecosystem, March 2019.

Chapter 4

Kwan, Savio. In-person conversation with author, December 28, 2021.

Ma, Jack. Speech to senior management, April 15, 2009, Alibaba headquarters, Hangzhou.

Puett, Michael, and Christine Gross-Loh. *The Path: What Chinese Philosophers Can Teach Us About the Good Life.* New York: Simon and Schuster, 2016.

Chapter 5

Kwan, Savio. In-person conversation with author, January 3, 2022.

Kwan, Savio. "Savio Kwan: The Alibaba Story." AGI lecture series, July 2, 2021.

Wong, Brian A. "Alibaba Leadership." Alibaba Business School, April 2019.

Wong, Brian A. "Road to 102: How Alibaba Works (Mission, Vision, Values)." AGI lecture series, March 22, 2022.

Chapter 6

Wong, Brian A. "Road to 102: How Alibaba Works (Strategy)." AGI lecture series, March 24, 2022.

Zeng, Ming. "My View on Strategy." AGI course material, July 5, 2021.

Zeng, Ming. *Smart Business: What Alibaba's Success Reveals About the Future of Strategy.* Boston: Harvard Business Review Press, 2018.

Chapter 7

Alibaba Group. "Corporate Governance: Alibaba Partnership." https:// alibabagroup.com/en/ir/governance_9.

Cai, Neo. "Road to 102: How Alibaba Works (Organization and People)." AGI lecture series, March 29, 2022.

Hu, Simon. DingTalk and in-person conversation with author, November 2021 and February 2022.

Jiang, Jane. DingTalk conversation and email correspondence with author, January 18, 2022.

Peng, Lucy. "The Heart, Head and Hand of an Organization." Alibaba Business School, May 7, 2021.

Wuxia Society. "Jin Yong Novels." https://wuxiasociety.com/jin-yong -novels/

Zeng, Ming. *Smart Business: What Alibaba's Success Reveals About the Future of Strategy*. Boston: Harvard Business Review Press, 2018.

Chapter 8

Barboza, David. "2 Executives Quit Alibaba.com Amid Fraud Inquiry." *New York Times*, February 21, 2011. www.nytimes.com/2011/02/22/business /global/22alibaba.html.

Cai, Neo. "Performance Management." Lecture, Alibaba Business School, April 2019.

Diabate, Dean H. Telephone conversation with author, October 22, 2021.

Epstein, Gady. "Alibaba's Jack Ma Fights to Win Back Trust." *Forbes*, March 23, 2011. www.forbes.com/forbes/2011/0411/features-jack-ma-alibaba -e-commerce-scandal-face-of-china.html?sh=61a74a6b4af5.

Greenleaf, Robert K. "The Servant as Leader." Robert K. Greenleaf Center for Servant Leadership, 2021. www.greenleaf.org/what-is-servant -leadership/.

Jing, Liu. "Alibaba Fires Employees for Mooncake Fraud." *China Daily*, September 14, 2016. www.chinadaily.com.cn/china/2016-09/14/content _26797048.htm.

Kwan, Savio. In-person conversation with author, December 28, 2021.

Xue, Yujie. "What Is China's 996 Work Culture That Is Polarising Its Silicon Valleys?" *South China Morning Post*, January 19, 2021. www.scmp .com/tech/tech-trends/article/3136510/what-996-gruelling-work-culture -polarising-chinas-silicon-valley.

Chapter 9

Alizila staff. "Daniel Zhang on Leadership and His Vision for Alibaba." Alizila, April 29, 2021. www.alizila.com/daniel-zhang-on-leadership-and -his-vision-for-alibaba/.

Business Roundtable. "Statement on the Purpose of a Corporation." Updated August 2019. https://opportunity.businessroundtable.org /ourcommitment/.

Collins, Jim. "Level 5 Leadership: The Triumph of Humility and Fierce Resolve." *Harvard Business Review*, January 1, 2001. https://hbr.org/2001 /01/level-5-leadership-the-triumph-of-humility-and-fierce-resolve-2.

Gonzales, Yuji Vincent. "Aquino Shares 'LQ' Conversation with Jack Ma." Inquirer.net, November 24, 2015. https://business.inquirer.net /203117/aquino-shares-lq-conversation-with-jack-ma.

Hupan University. "How to Make Decisions as a Leader." AGI Course Reading, May 7, 2021. www.yuque.com/agi_insights/how-alibaba-works /make-decision-as-leader.

Jing, Eric. "Eric Jing on the Promise of Financial Services for the Unbanked." *Wall Street Journal*, January 17, 2018. www.wsj.com/articles/eric-jing-on -the-promise-of-financial-services-for-the-unbanked-1516200702.

Lao Tzu. *Tao Te Ching*. Translated by Arthur Waley. New York: Wordsworth Editions, 1996.

McChrystal, Stanley. *Team of Teams: New Rules of Engagement for a Complex World*. New York: Portfolio, 2015.

Schwab, Klaus. "Why We Need the 'Davos Manifesto' for a Better Kind of Capitalism." World Economic Forum, December 1, 2019. www.weforum .org/agenda/2019/12/why-we-need-the-davos-manifesto-for-better-kind -of-capitalism/.

Schwantes, Marcel. "Google's Insane Approach to Management Could Transform Your Company." *Inc.*, November 22, 2016. www.inc.com/marcel -schwantes/googles-insane-approach-to-management-could-transform -your-company.html.

Schwantes, Marcel. "Self-Made Billionaire Jack Ma Says You'll Need This 1 Rare Skill to Succeed in the Age of Machines." *Inc.*, October 5, 2017. www.inc.com/marcel-schwantes/1-rare-trait-that-actually-trumps-iq-emo tional-intelligence-says-billionaire-jack-ma.html.

Wojcicki, Esther. Zoom conversation with author, January 11, 2022.

Wong, Brian A. "Alibaba Leadership." Alibaba Business School, April 2019.

World Economic Forum. "Davos 2019—Meet the Leader with Alibaba Executive Chairman Jack Ma." Interview by Olajumoke Adekeye, February 10, 2019. YouTube video, 55:59. www.youtube.com/watch?v=ZvwBVllprWY.

World Economic Forum. "Jack Ma: Love Is Important in Business, Davos 2018." Interview by Abi Ramana, January 24, 2018. YouTube video, 55:39. www.youtube.com/watch?v=4zzVjonyHcQ.

Xiu, Ouyang. *Historical Records of the Five Dynasties*. Translated by Richard L. Davis. New York: Columbia University Press, 2008.

Chapter 10

Alibaba. "Alibaba Group's Innovative Recruiting Program Supports International Growth." October 24, 2016. www.alibabagroup.com/en/news /article?news=p161024.

Aliresearch. "Digital Economy Revitalizes Rural China." January 17, 2021. www.aliresearch.com/en/Reports/Reportsdetails?articleCode=21855.

Alizila staff. "An Introduction to Taobao Villages." Alizila, January 18, 2016. www.alizila.com/an-introduction-to-taobao-villages/.

Business Wire. "MYbank's Use of Digital Technology Leads to Record Growth in Rural Clients." Bloomberg, June 23, 2021. www.bloomberg.com/press-releases/2021-06-24/mybank-s-use-of-digital-technology-leads-to-record-growth-in-rural-clients.

CGAP. "China: A Digital Payments Revolution." Consultative Group to Assist the Poor, September 2019. www.cgap.org/research/publication/china-digital-payments-revolution.

China Daily staff. "Inside a Taobao Village." China Daily, February 27, 2015. www.chinadaily.com.cn/business/2015-02/27/content_19670330_6.htm.

Flood, Ethan Cramer. "In Global Historic First, E-Commerce in China Will Account for More Than 50% of Retail Sales." eMarketer, February 10, 2021. www.emarketer.com/content/global-historic-first-ecommerce-china-will-account-more-than-50-of-retail-sales.

Han, Sun. In-person conversation with author in Shaji township, April 9, 2013.

Hofman, Bert. "The Taobao Villages as an Instrument for Poverty Reduction and Shared Prosperity." World Bank Blogs, October 29, 2016. www.worldbank.org/en/news/speech/2016/10/29/the-taobao-villages-as-an-instrument-for-poverty-reduction-and-shared-prosperity.

Liu, Shumin. "Dive into the Poverty Line to Foster Inclusive Growth in Rural China." Regional Innovation Centre UNDP Asia-Pacific, November 20, 2019. https://undp-ric.medium.com/sink-in-the-poverty-line-to-foster-inclusive-growth-with-villages-in-china-2ec8a426740e.

Luo, Xubei. "E-Commerce for Poverty Alleviation in Rural China: From Grassroots Development to Public-Private Partnerships." World Bank Blogs, March 19, 2019. https://blogs.worldbank.org/eastasiapacific/e-commerce-poverty-alleviation-rural-china-grassroots-development-public-private-partnerships.

Luo, Xubei, and Chiyu Niu. "E-Commerce Participation and Household Income Growth in Taobao Villages." Poverty Equity Global Practice Working Paper 198, World Bank Group, April, 2019. https://documents1.worldbank.org/curated/en/839451555093213522/pdf/E-Commerce-Participation-and-Household-Income-Growth-in-Taobao-Villages.pdf.

Luohan Academy. Digital Technology and Inclusive Growth. Luohan Academy report, 2019.

People's Bank of China. "中国普惠金融指标分析报告." 中国人民银行金融消费权益保护局, 2020. http://www.pbc.gov.cn/goutongjiaoliu/113456/113469/4335821/2021090816343161697.pdf.

Thoreau, Henry David. *Walden*. New York: Empire Books, 2012.

Wong, Brian A. "Can China Balance Consumerism with Sustainability?" Alizila, November 12, 2019. www.alizila.com/can-china-balance-consumerism-with-sustainability/.

World Bank. "Stimulating Jobs, Growth, Entrepreneurship, Income in Rural China through E-Commerce." November 22, 2019. www.worldbank.org/en/results/2019/11/22/stimulating-jobs-growth-entrepreneurship-income-in-rural-china-through-e-commerce.

Wu, Julia. "A Brief History of Jack Ma's Ant Financial—the $150B Unicorn." Hacker Noon, August 6, 2019. https://hackernoon.com/the-story-of-ant-financial-4t2aq3zh8.

Xinhua. "China's Non-Cash Payments Top 4,000 Trillion Yuan in 2020." March 27, 2021. http://www.xinhuanet.com/english/2021-03/27/c_139840050.htm.

Xu, Mengqi. "How Internet Connectivity Has Changed People's Lives in China." CGTN, October 23, 2019. https://news.cgtn.com/news/2019-10-22/How-internet-connectivity-has-changed-people-s-lives-in-China-KZR0tk5tra/index.html.

Zeng, Ming. "Alibaba and the Future of Business: Lessons from China's Innovative Digital Giant." *Harvard Business Review*, September–October 2018. https://hbr.org/2018/09/alibaba-and-the-future-of-business.

Chapter 11

Alibaba Global Initiatives. "Our Mission." https://agi.alibaba.com/.

Azmi, Dzof. "DESA: Helping Rural Malaysian Farmers Reap the Rewards of E-Commerce." Digital News Asia, December 12, 2019. www.digitalnewsasia.com/digital-economy/desa-helping-rural-malaysian-farmers-reap-rewards-e-commerce.

Cameroon Online staff. "Alibaba Business School and UNCTAD Train African Entrepreneurs to Become Catalysts for Digital Transformation." Cameroon Online News, July 2, 2018, https://cameroononlinenews.com/alibaba-business-school-and-unctad-train-african-entrepreneurs-to-become-catalysts-for-digital-transformation/.

Gilchrist, Karen. "This 32-Year-Old's Start-Up Is Helping Thousands of Malaysians Find Work During the Pandemic." Make It, CNBC, October 28, 2020. www.cnbc.com/2020/10/29/this-helping-during-covid.html.

Hull, Callum. "CEO Interview: Glints—Empowering People and Organization." HR Digital Today, August 6, 2020. www.hrdigitaltoday.com /blog/2020/08/ceo-interview-glints-empowering-peopleorganizations.

Iderawumi, Mustapha. "In Conversation with Clarisse Iribagiza, CEO of HeHe." Space in Africa, June 7, 2021. https://africanews.space/in -conversation-with-clarisse-iribagiza-ceo-of-hehe/.

Iribagiza, Clarisse. Email and messenger correspondence with author, September 2021.

Ismail, Izwan. "Moving to Cloud Helps Businesses Operate During Pandemic." New Straits Times, July 22, 2021. www.nst.com.my/lifestyle /bots/2021/07/710603/tech-moving-cloud-helps-businesses-operate-during -pandemic.

Kene-Okafor, Tage. "African Fintech Flutterwave Triples Valuation to over $3B After $250 Million Series D." TechCrunch, February 16, 2022. https://techcrunch.com/2022/02/16/african-fintech-flutterwave-triples -valuation-to-over-3b-after-250m-series-d/.

Kim, Jay. "This Malaysian Start-Up Can Help You with Your To-Do List, On Demand." Tech in Asia, July 16, 2018. www.techinasia.com/talk /francesca-chia-cofounder-ceo-goget.

Leong, Clarence. Email and video conference conversations with author, October 2021.

Limos, Mario Alvaro. " 'Salamat sa Ani' Lets You Help Filipino Farmers Conquer the Pandemic." Esquire, March 16, 2021. www.esquiremag.ph /long-reads/features/salamat-sa-ani-a00293-20210316-lfrm.

Namyalo, Consolate. "Clarisse Iribagiza Is Transforming Rwanda into one of Africa's Most Active Tech Hubs." Glim, July 11, 2019. https:// glimug.com/clarisse-iribagiza-is-transforming-rwanda-into-one-of-africas -most-active-tech-hubs/.

Tsai, Joseph. Alibaba international staff communication re: Alibaba Group Mission, April 23, 2010.

UNCTAD. "UNCTAD and Alibaba Business School Kick Off eFounders Fellowship for Asian E-Commerce Entrepreneurs." United Nations Conference on Trade and Development, March 26, 2018. https://unctad .org/news/unctad-and-alibaba-business-school-kick-efounders-fellowship -asian-e-commerce-entrepreneurs.

UNCTAD. "Young Digital Entrepreneurs Leading Africa into a New Era." United Nations Conference on Trade and Development, November 28, 2019. https://unctad.org/news/young-digital-entrepreneurs-leading-africa-new-era.

Uzo-Ojinnaka, Uju. Speech, African Heroes Conference, Accra, Ghana, August 2019.

Wong, Brian A. "How Digital Entrepreneurs Are Accelerating Economic Recovery After COVID-19 Pandemic." Thoughts, June 29, 2020. www.bernama.com/en/thoughts/news.php?id=1854262.

Yeo, Oswald. Email and video conference conversations with author, September 2021.

Yousuf, Faiza. "Yanum Kamran—Incredible #WomenInTechPK!" WomenInTechPK, July 8, 2019. www.womenintechpk.com/anum-kamran-in credible-women-in-tech/.

Yun, Tan Zhai. "Logistics: Can Pgeon Crack the Courier Market?" The Edge Malaysia, The Edge Markets, November 29, 2021. www.theedgemarkets .com/article/logistics-can-pgeon-crack-courier-market%C2%A0.

Chapter 12

Andreessen Horowitz. *How to Win the Future: An Agenda for the Third Generation of the Internet.* October 2021. https://a16z.com/wp-content /uploads/2021/10/How-to-Win-the-Future-1.pdf.

Fuller, R. Buckminster. *Operating Manual for Spaceship Earth.* Carbondale: Southern Illinois University Press, 1969.

Fung, Yu-lan. *A Short History of Chinese Philosophy.* Edited by Derk Bodde. New York: Macmillan, 1948.

Hoff, Benjamin. *The Tao of Pooh.* New York: Penguin Books, 1982.

Lebow, Sara. "Proximity Mobile Payments Are Massively Popular in China, but Usage Lags in U.S." Inside Intelligence, eMarketer, July 19, 2021. www.emarketer.com/content/proximity-mobile-payments-massively -popular-china-usage-lags-us.

Marr, Bernard. *Big Data in Practice: How 45 Successful Companies Used Big Data Analytics to Deliver Extraordinary Results.* Hoboken, NJ: John Wiley and Sons, 2016.

Puett, Michael, and Christine Gross-Loh. *The Path: What Chinese Philosophers Can Teach Us About the Good Life.* New York: Simon and Schuster, 2016.

Turner, Fred. "R. Buckminster Fuller: A Technocrat for the Counterculture." Chapter 9 in *New Views on R. Buckminster Fuller,* edited by Hsiao-Yun Chu and Robert G. Trujillo. Redwood City, CA: Stanford University Press, 2009.

World Bank. "Overview: The World Bank in China." October 21, 2021. www.worldbank.org/en/country/china/overview#3.

World Bank. "Poverty Headcount Ratio at $1.90 a Day (2011 PPP) (% of population)—China." Graph, 2011–2018. https://data.worldbank.org /indicator/SI.POV.DDAY?locations=CN.

Index

© THANAKRIT GU

BRIAN A. WONG is a Chinese American entrepreneur and investor. He was the first American and only the fifty-second employee to join Alibaba Group, where he contributed to the company's early globalization efforts and served as Jack Ma's special assistant for international affairs. During his sixteen-year tenure, Wong established the Alibaba Global Initiatives (AGI) division and was the founder and executive director of the Alibaba Global Leadership Academy. Wong remains an adviser to the AGI team and regularly teaches courses on China's digital economy and the Tao of Alibaba management principles. Wong founded RADII, a digital media platform dedicated to bridging understanding between East and West.

Wong earned his bachelor's degree from Swarthmore College, a master's certificate from the Johns Hopkins University (SAIS)–Nanjing University Center for US and China Studies, and an MBA from the University of Pennsylvania's Wharton School. He was selected as a Young Global Leader by the World Economic Forum in 2015 and is a China Fellow with Aspen Institute, a member of the Aspen Global Leadership Network, and a member of the Committee of 100. He is based in Shanghai, China.

PublicAffairs is a publishing house founded in 1997. It is a tribute to the standards, values, and flair of three persons who have served as mentors to countless reporters, writers, editors, and book people of all kinds, including me.

I. F. STONE, proprietor of *I. F. Stone's Weekly*, combined a commitment to the First Amendment with entrepreneurial zeal and reporting skill and became one of the great independent journalists in American history. At the age of eighty, Izzy published *The Trial of Socrates*, which was a national bestseller. He wrote the book after he taught himself ancient Greek.

BENJAMIN C. BRADLEE was for nearly thirty years the charismatic editorial leader of *The Washington Post*. It was Ben who gave the *Post* the range and courage to pursue such historic issues as Watergate. He supported his reporters with a tenacity that made them fearless and it is no accident that so many became authors of influential, best-selling books.

ROBERT L. BERNSTEIN, the chief executive of Random House for more than a quarter century, guided one of the nation's premier publishing houses. Bob was personally responsible for many books of political dissent and argument that challenged tyranny around the globe. He is also the founder and longtime chair of Human Rights Watch, one of the most respected human rights organizations in the world.

. . .

For fifty years, the banner of Public Affairs Press was carried by its owner Morris B. Schnapper, who published Gandhi, Nasser, Toynbee, Truman, and about 1,500 other authors. In 1983, Schnapper was described by *The Washington Post* as "a redoubtable gadfly." His legacy will endure in the books to come.

Peter Osnos, *Founder*